T0215324

Estimands, Estimators and Sensitivity Analysis in Clinical Trials

Chapman & Hall/CRC Biostatistics Series

Series Editors

Shein-Chung Chow, Duke University School of Medicine, USA

Byron Jones, Novartis Pharma AG, Switzerland

Jen-pei Liu, National Taiwan University, Taiwan

Karl E. Peace, Georgia Southern University, USA

Bruce W. Turnbull, Cornell University, USA

Estimands, Estimators and Sensitivity Analysis in Clinical Trials

Craig Mallinckrodt

Geert Molenberghs

Ilya Lipkovich

Bohdana Ratitch

CRC Press
Taylor & Francis Group
Boca Raton London New York

CRC Press is an imprint of the
Taylor & Francis Group, an **informa** business

A CHAPMAN & HALL BOOK

CRC Press
Taylor & Francis Group
6000 Broken Sound Parkway NW, Suite 300
Boca Raton, FL 33487-2742

First issued in paperback 2021

© 2020 by Taylor & Francis Group, LLC
CRC Press is an imprint of Taylor & Francis Group, an Informa business

No claim to original U.S. Government works

ISBN 13: 978-1-03-224262-0 (pbk)
ISBN 13: 978-1-138-59250-6 (hbk)

DOI: 10.1201/9780429488825

This book contains information obtained from authentic and highly regarded sources. Reasonable efforts have been made to publish reliable data and information, but the author and publisher cannot assume responsibility for the validity of all materials or the consequences of their use. The authors and publishers have attempted to trace the copyright holders of all material reproduced in this publication and apologize to copyright holders if permission to publish in this form has not been obtained. If any copyright material has not been acknowledged please write and let us know so we may rectify in any future reprint.

Except as permitted under U.S. Copyright Law, no part of this book may be reprinted, reproduced, transmitted, or utilized in any form by any electronic, mechanical, or other means, now known or hereafter invented, including photocopying, microfilming, and recording, or in any information storage or retrieval system, without written permission from the publishers.

For permission to photocopy or use material electronically from this work, please access www.copyright. com (http://www.copyright.com/) or contact the Copyright Clearance Center, Inc. (CCC), 222 Rosewood Drive, Danvers, MA 01923, 978-750-8400. CCC is a not-for-profit organization that provides licenses and registration for a variety of users. For organizations that have been granted a photocopy license by the CCC, a separate system of payment has been arranged.

Trademark Notice: Product or corporate names may be trademarks or registered trademarks, and are used only for identification and explanation without intent to infringe.

Publisher's Note
The publisher has gone to great lengths to ensure the quality of this reprint but points out that some imperfections in the original copies may be apparent.

Visit the Taylor & Francis Web site at
http://www.taylorandfrancis.com

and the CRC Press Web site at
http://www.crcpress.com

Contents

Section I   Setting the Stage

Section II   Estimands

Section III   Estimators and Sensitivity

Section IV   Technical Details on Selected Analyses

Section V   Case Studies: Detailed Analytic Examples

List of Figures

List of Tables

List of Code Fragments

Preface

The ICH E9(R1) Addendum provides not only a new view on analytic approaches but also a new approach to study development. This new focus on estimands has revealed challenges and previously masked problems for clinical statisticians. This shift in focus requires rethinking the way we consider clinical objectives, missing data, and choosing appropriate estimators. The new estimand framework also has revealed gaps between different communities such as causal inference and clinical trialists and in some instances a lack of a common language to consider key issues. The aim of the addendum is to make treatment effects easier to interpret; statisticians must be prepared for the new analytic and cross-functional challenges raised by the addendum. We do not claim that this book solves all problems, nor does it cover all issues. However, we provide a unifying framework along with the technical details and practical guidance to help statisticians meet these challenges.

Specifically, the ICH E9(R1) Addendum focuses on issues, and how to deal with them, that arise from intercurrent events (ICEs), which are post-randomization events that are related to treatment and/or outcome. These events can make it difficult to understand what we should try to estimate from clinical trial data and how we can estimate those quantities. This book provides details on how clear estimands lay a foundation for specifying aspects of trial design, conduct, and analysis, needed to overcome the issues caused by ICEs. A study development process chart is presented as a tool to guide choice of objectives, estimands, study design, and analysis. The role of the intention-to-treat principle is discussed in this new framework. Emphasis is placed on defining the treatment or treatment regimen of interest and to define which ICEs are a deviation from the regimen of interest. The five strategies for dealing with ICEs noted in the ICH E9(R1) Addendum are examined in detail, including the assumptions, implications, strengths, and limitations for each strategy. Examples of choosing and specifying estimands in real-life clinical trial scenarios are provided. Classic clinical trial concepts are linked with corresponding concepts and ideas from causal inference.

The desired attributes of an estimator in the framework outlined in the ICH E9(R1) Addendum are discussed along with the general modeling considerations for longitudinal clinical trial data. Overviews of specific analytic methods and analytic considerations for each of the five strategies for dealing with ICEs that are outlined in the ICH Addendum are discussed. An overview of the most recent and traditional concepts on handling missing data is discussed along with an overview of sensitivity analyses. Technical details on analytic approaches that can be used as the main estimator and sensitivity analyses consistent with each of the five strategies for dealing with ICEs as outlined in the ICH E9(R1) Addendum are presented. An example data set is used, and the example code is provided to implement each of these analyses.

The book closes with case studies from diverse settings that illustrate and reinforce many of the concepts presented throughout this book. Our hope and belief are that statisticians who rigorously study the concepts and practical approaches to implement those concepts presented in this book will be well prepared to lead teams in the study development process and to implement analytic plans that will meet the aim of the ICH E9(R1) Addendum of making treatment effects easier to understand.

Acknowledgments

We thank our colleagues and friends for many discussions that inspired and informed this book. We specifically thank the Drug Information Association Scientific Working Group (DIASWG) on missing data. As members of this group, we benefitted significantly from the group's many discussions and from our individual discussions with other group members. In this book, we have frequently cited work of the group's members. Most notably, the three papers listed below were developed within the DIASWG and are the basis for multiple chapters in this book. As relevant, individual authors from these papers are cited as having contributed content within chapters of this book.

Ratitch B, Bell J, Mallinckrodt C, Bartlett J, Goel N, Molenberghs G, O'Kelly M, Singh P, Lipkovich I. Choosing Estimands in Clinical Trials: Putting the ICH E9(R1) into Practice. Therapeutic Innovation and Regulatory Science. 2020. 54(2):324–341.

Mallinckrodt C, Bell J, Liu G, Ratitch B, O'Kelly M, Lipkovich I, Singh P, Xu L, Molenberghs G. Technical and Practical Considerations in Aligning Estimators with Estimands in Clinical Trials. Therapeutic Innovation and Regulatory Science. 2020 54(2):353–364.

Ratitch B, Goel N, Mallinckrodt C, Bell J, Bartlett J, Molenberghs G, Singh P, Lipkovich I, O'Kelly M. Defining Estimands in Clinical Trials: Examples Illustrating ICH E9(R1) Guidelines. Therapeutic Innovation and Regulatory Science. 54(2):370–384.

Authors

Craig Mallinckrodt holds the rank of distinguished biostatistician at Biogen in Cambridge, Massachusetts. He has extensive experience in all phases of clinical research. His methodology research focuses on longitudinal and incomplete data. He is a fellow of the American Statistical Association, has led several industry working groups on missing and longitudinal data, and received the Royal Statistical Society's award for outstanding contribution to the pharmaceutical industry.

Geert Molenberghs is a professor of biostatistics (Hasselt University, KU Leuven). He works on surrogate endpoints, longitudinal and incomplete data, was an editor for *Applied Statistics*, *Biometrics*, *Biostatistics*, *Wiley Probability & Statistics*, and *Wiley StatsRef*, and is an executive editor of *Biometrics*. He was the president of the International Biometric Society, is a fellow of the American Statistical Association, and received the Guy Medal in Bronze from the Royal Statistical Society. He has held visiting positions at the Harvard School of Public Health.

Ilya Lipkovich is a senior research advisor at Eli Lilly and Company. He is a fellow of the American Statistical Association and has published on subgroup identification in clinical data, analysis with missing data, and causal inference. He is a frequent presenter at conferences, a co-developer of subgroup identification methods, and a co-author of the book *Analyzing Longitudinal Clinical Trial Data: A Practical Guide*.

Bohdana Ratitch is a principal research scientist at Eli Lilly and Company. Bohdana has contributed to research and practical applications of methodologies for causal inference and missing data in clinical trials through active participation in a pharma industry working group, numerous publications, presentations, and co-authoring the book *Clinical Trials with Missing Data: A Guide for Practitioners*.

Section I

Setting the Stage

Section I begins with an introductory chapter that describes the basic problem of how post-randomization events that are related to both treatment and outcome can bias estimates of treatment effects and thereby cloud inferences. The opening chapter also describes some of the history of how these issues were statistically dealt with in the narrow context of missing data and begins the transition to the newer ways of conceptualizing the problem, especially as outlined in the ICH E9(R1) Addendum on estimands and sensitivity analyses. The second chapter sets the stage for Section II by focusing on why estimands are important.

1

Introduction

1.1 Understanding the Problem

A simple definition of the term "estimand" is what is to be estimated. As seen in subsequent chapters, choosing, specifying, and properly evaluating various estimands is a nuanced, detailed, and complex topic. Drug development has proceeded for many decades with clinical trials that specified general, primary, and secondary objectives but did not clearly specify estimands. So why do we need estimands and all the complexity that comes with them? Is this just another regulatory requirement?

Yes, clearly specifying estimands is a regulatory requirement in the ICH E9(R1) Addendum (ICH, 2019). However, clearly specifying estimands is essential because they lay the foundation for understanding the treatment effects. We can improve drug development, clarify and simplify analysis plans, and more clearly characterize the benefits and risks of potential medicines if we utilize clearly specified estimands.

To motivate discussion, the data in Figure 1.1 are taken, which depicts a randomized clinical trial comparing an experimental treatment to a control based on mean change from baseline to endpoint. In these data, the treatment effect is clear. The experimental group has an advantage of four points over the control group. Similarly, there is little controversy about appropriate analytic approaches. Although, depending on certain conditions, debate about parametric versus nonparametric methods, or considerations about covariate adjustment might arise, simple, easy-to-understand analyses will yield a valid estimate of the treatment effect. Unfortunately, data such as that shown in Figure 1.1 is not realistic in most clinical trial scenarios.

In the real world, data are like those depicted in Figure 1.2. Patients discontinue treatments at different times, for different reasons, and at different rates; they may or may not take a subsequent treatment and may or may not continue to be evaluated in the trial. These treatment-related post-randomization events create ambiguity not just about how to analyze the

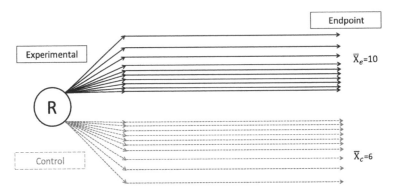

FIGURE 1.1
Data depiction from an ideal clinical trial.

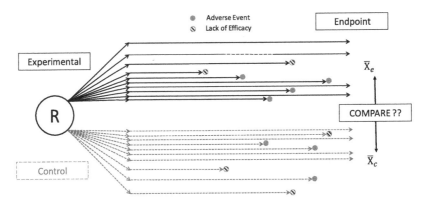

FIGURE 1.2
Data depiction from a realistic clinical trial.

data but more fundamentally about what to compare. For example, can we, or should we, still focus on mean change to the planned endpoint of the trial? Or, should we use knowledge of the post-randomization events to estimate some other aspects of response to treatment? Or, is it the presence of a post-randomization event itself that is most meaningful?

For many years, statisticians and clinical trialists considered various ways for HOW to estimate certain parameters from data such as those in Figure 1.2, implying that the most relevant question was still to assess mean change to the originally planned endpoint. More recently, we have begun to focus first on WHAT we should try to estimate, and only then we can consider how that can be done in the most valid and least biased manner.

In this book, we discuss both what we should try to estimate in clinical trial data and how we can estimate those quantities.

1.2 History

Understanding how to deal with data such as those depicted in Figure 1.2 entails much more than missing data considerations. However, to understand the history and how we arrived at our current state, it is useful to start with a focus on missing data because, historically, that was the focus of discussion. Until recently, guidelines for the analysis of clinical trial data provided only limited advice on how to handle missing data and other post-randomization events that could be related to treatment and outcome. Analytic approaches tended to be simple and ad hoc (Molenberghs and Kenward, 2007). These methods were motivated by limited computing power and restored the intended balance to the data, allowing implementation of the simple analyses for complete data.

The seminal work on analysis of incomplete data by Rubin (1976) began to shift attention from merely restoring balance for the sake of computational ease to accounting for the potential bias from incomplete data. A core feature of Rubin's (1976) work included the now common taxonomy of missingness mechanisms: missing completely at random (MCAR), missing at random (MAR), and missing not at random (MNAR) (see Chapter 15 for more details).

Despite these advances, widespread use of simple methods for dealing with incomplete data set historical precedent and fostered their continued acceptance in regulatory settings even after alternative methods were well known. In regulatory settings, methods such as baseline and last observation carried forward remained popular choices for the primary analysis long after more principled alternatives could be easily implemented with commercial software (Mallinckrodt, 2013).

In addition to historical precedent, continued acceptance of simple methods despite strong evidence of their shortcomings was also fostered by the belief that these methods led to conservative estimates of treatment effects. Conservative in this context meant not favoring the experimental treatment, that is, underestimating the advantage of experimental treatment over control. Hence, even though it became well known that simple methods often yielded biased estimates of treatment effects, that bias was considered appropriate for confirmatory clinical trials of medical interventions because the bias was thought to provide protection against erroneous approval of ineffective interventions (Kenward and Molenberghs, 2009).

However, widespread research in the 1990s and 2000s from independent investigators across academia, industry, and regulatory authorities showed that conditions under which simple methods yield conservative estimates and maintain control of Type I error are not straightforward and cannot be assured at the start of the trial (Mallinckrodt, 2013). Moreover, newer, more principled methods such as multiple imputation and likelihood-based repeated measure analyses were shown to have less bias or no bias in situations where simple methods performed poorly (Mallinckrodt, 2013).

Not surprisingly, choice of the primary analysis became a frequent topic at academic conferences and in regulatory discussions.

This research and the associated discussions set the stage for new and updated guidance for preventing and handling missing data in clinical trials. Most importantly, an expert panel from the National Research Council (NRC) that was commissioned by the Food and Drug Administration (FDA) issued an extensive set of recommendations (NRC, 2010). Although the panel focused on phase III trials, the recommendations are useful regardless of the stage of development. The recommendations set forth an overarching framework for tackling the problem of missing data. Key pillars of this framework include (1) providing precise and clear specification of trial objectives and the associated estimands, (2) designing and conducting trials in a way to maximize adherence to the protocol-defined interventions, and (3) including a sensible primary analysis and sensitivity analyses that support the research question and assess robustness of the primary result to missing data assumptions.

After the NRC report, estimands became a focal point for research and discussion in drug development. The report stressed that estimands, the population quantity to be estimated to address the objectives of a study, should be defined first, and then analyses should be chosen in alignment with the estimand. A series of publications led to a process chart for choosing estimands (Mallinckrodt et al., 2012; Leuchs et al., 2015; Garrett, 2015; Phillips et al., 2016). However, the NRC report focused on missing data, and hence it was not surprising that subsequent discussions about estimands continued to focus on missing data and the analytic methods for dealing with them. In practice, choice of estimands and the associated inferences and interpretations still tended to be driven by habitual choices of analytical methods from which the choice of estimand was implicitly inferred (Ratitch et al., 2019a).

To illustrate the consequences that failure to clearly identify a primary estimand can have, we can consider how the intention-to-treat (ITT) principle has been applied historically. The original ICH E9 guidance (ICH, 1998) has been the foundation for statistical considerations in the design, analysis, and reporting of clinical trials. Although the guidance focused on confirmatory trials, its core principles of seeking to minimize bias are useful in every phase of the clinical development. For example, blinding and randomization are standard design features that we almost take for granted now in eliminating assessment and selection bias. The ICH E9 guidance also laid out the fundamental tenants of the ITT principle, describing it as "the principle that asserts that the effect of a treatment policy can be best assessed by evaluations based on the intention to treat a patient (i.e., the planned treatment regimen) rather than the actual treatment given."

The ITT principle has two components: what patients to include in the analysis and what data on each patient to include. The original aim of the ITT principle was primarily on what patients to include. The ITT principle recognized the need to avoid excluding patients who did not adhere to the planned treatment regimen to preserve the causal link to randomization and to reduce bias that could result from using a selected subset of patients. As such, ITT became the foundation for defining analysis sets, with primary inferences typically driven by analyses based on all randomized patients. However, in retrospect, it is compelling to note that the notion of estimating the effects of being randomized to a treatment, versus say the effects of the initially randomized treatment, was not discussed or justified.

Following a strict interpretation of ITT requires the collection and analysis of all data regardless of adherence to the randomized study treatments. Historically, postdeviation data was not always collected. Sometimes, postdeviation data were collected but not included in the analysis. The motivations for the various approaches were seldom clear at the time, but in retrospect seem likely driven by implicit focus on differing estimands. Some may have considered postdeviation data to be uninformative for the treatment effect of interest; others may have simply followed the established practice at the time without thinking about the implications.

Regardless of the motivation that led to this practice, not collecting all relevant data or excluding such data from the statistical analysis had for some time been framed strictly as a missing data problem. At the same time, most were claiming to conduct analyses in accordance with the ITT principle by virtue of including all randomized patients in the analysis set while using various methods of imputation or implicit assumptions about the missing postdeviation data. The implications of such imputations or assumptions on the interpretation of the estimated treatment effect were rarely discussed and fully appreciated. Of course, following the strict ITT principle of including all outcomes as planned regardless of post-randomization events reduces the amount of missing data compared to not including data after post-randomization events. However, the effect estimated under the strict ITT principle in the presence of post-randomization nonadherence represents the effect of being randomized to a treatment and may not always be attributable to the randomized treatment itself. Such an estimand, sometimes referred to as a pure treatment policy estimand, may not always be appropriate for decision making by various stakeholders. Therefore, simply reducing the amount of missing data is no justification for an estimand (Fleming, 2011). The choice of estimand should be driven by the scientific question, not ease of analysis (Mallinckrodt et al., 2017; Akacha et al., 2017).

For decades, clinical scientists, statisticians, and regulators discussed methods to handle the missing data problem. Without a clear understanding of the estimand, sometimes the scientists lost sight of relevant questions such as: Why are the data missing? Could they have been collected, and if so, would they be meaningful? Does the event leading to missing data represent an important clinical outcome in itself? For example, is it meaningful to consider a symptom score or quality-of-life assessment to be missing after a patient died? In this case, the data are incomplete compared to what was intended, but are we missing meaningful information? Alternatively, is it clinically meaningful to use symptom scores that were collected after a patient discontinued the study treatment and/or started taking an alternative medication? Or, is the fact that the patient was not able to tolerate or for whatever reason unable or unwilling to continue the treatment the main outcome of interest?

The answers to these questions depend on the context and the specific aim of the trial, along with the therapeutic area, and how the treatment benefit is to be assessed – and on what is to be estimated – the estimand. In other words, it depends on what it is that we want to know and estimate. Well-defined estimands make it clear what data are needed and what analyses are appropriate.

The recent ICH E9(R1) Addendum on "Estimands and Sensitivity Analysis in Clinical Trials" (ICH, 2019) expands the discussion beyond missing data to provide a broader view of post-randomization events that may be related to treatment and outcome (intercurrent events). The addendum provides a language and framework for choosing estimands and reverses the habitual practice of analytic choices implicitly defining objectives and estimands (Ratitch et al., 2019a).

The intent of this book is to provide practical guidance in implementing the latest research and guidance for choosing estimands, estimators, and sensitivity analyses in clinical trials.

1.3 Summary

In the analysis of clinical trial data, problems arise from post-randomization events that are related to treatment and outcome. These events can make it difficult to understand what we should try to estimate in clinical trial data and how we can estimate those quantities. Clinical trial analyses have evolved from an early focus on simple methods of imputation of data that may be incomplete due to these post-randomization events in order to facilitate ease of computation and then shifting to more principled methods. The NRC expert panel report (NRC, 2010) initiated a shift in focus from merely how we should estimate to what we should estimate. The 2019 ICH E9(R1) Addendum on "Estimands and Sensitivity Analysis in Clinical Trials" (ICH, 2019) expanded analytic discussions beyond missing data to tackle the issues caused by the wider variety of post-randomization events that are related to the treatment and the outcome, referred to as intercurrent events.

2

Why Are Estimands Important?

2.1 Motivating Example

To further motivate the use of clearly defined estimands, consider the following example taken from Akacha et al. (2017). This example, from an FDA public advisory committee meeting (FDA, 2014a, 2014b, 2014c), illustrates in a regulatory submission context how the measure of treatment benefit left room for ambiguity in understanding the treatment effects of the antidiabetic drug dapagliflozin.

The primary endpoint in the pivotal trials was the change in HbA1c from baseline to 24 weeks. Throughout the trials, patients received glucose-lowering rescue medications for the remainder of the trial, if one of several markers of glycemic control exceeded prespecified thresholds. Data after initiation of rescue medication was considered missing. This appears to be common practice in the analysis of diabetes trials because use of rescue medication may mask or exaggerate the risks and benefits of the treatment, especially because use of rescue will occur more frequently in the less effective treatment group. Following standard practice at the time, the last prerescue medication value was carried forward and analyzed as the end-of-trial value.

The statistical reviewer at the FDA commented on the primary analysis using the last observation carried forward (LOCF) approach as follows (FDA, 2014c):

> While FDA has implicitly endorsed LOCF imputation for diabetes trials in the past, there is now more awareness in the statistical community of the limitations of this approach. In particular, the argument has been made that LOCF can be anti-conservative (i.e., it sometimes favors the alternative hypothesis more than other approaches) and the findings from the placebo-controlled studies that I reviewed bear this out.
>
> My own preferred analysis simply uses the observed values of patients who were rescued. This approach may seem counterintuitive if one believes that rescue treatment makes the subsequent outcomes less relevant to evaluation of the test agent. It has the virtue, however, of respecting the intent-to-treat principle, in the sense that the analysis is based on the randomized treatment rather than the treatment actually received (i.e., planned treatment plus rescue).

Akacha et al. (2017) note that at first glance it seems that both parties simply used a different statistical approach to analyze the data to reduce bias while targeting the same estimand. The sponsor removed the data after initiation of rescue medication and applied the LOCF approach to impute the resulting missing data. The FDA statistician refrained from using LOCF due to its poor statistical properties and included all data regardless of initiation of rescue medication. Even so, there must have been some missing values, for example, from patients that discontinued the trial. It is not clear from this description how FDA handled those missing values.

Importantly, the implications of using different statistical approaches to analyze the data go far beyond analytic implications. The implied scientific questions of interest that are answered by the two parties are fundamentally different. This important distinction was not clear in the FDA Advisory Committee proceedings. The sponsor seemed to be interested in establishing the effect of the initially randomized treatments had no patient received rescue medication, while the FDA statistician was comparing the treatment policies "dapagliflozin plus rescue" versus "control plus rescue."

The dapagliflozin example shows how differences in the scientific question arise due to different ways of handling data after treatment discontinuation or use of rescue medication. However, these differences were not clear to either the trialists or the statisticians from the sponsor or the regulators. Indeed, the intricate mathematics inherent to estimating treatment effects can be lost on all but the most knowledgeable statisticians and engender confusion and lack of interpretability (Akacha et al., 2017).

What has happened in this example and has been common in medical research is a backward approach in which the choice of analysis drives the scientific question that is addressed rather than the scientific question driving the analysis. Statisticians, trialists, and regulators discuss what statistical assumptions and methods to use to account for missing data (or other post-randomization events) without understanding and agreement on the scientific question or objective. Not surprisingly, this has led to disagreement about the plausibility of statistical assumptions, the merits of various analyses, and most importantly, the risks and benefits of the interventions under evaluation.

2.2 How Estimands Help

As discussed in Chapter 3, clearly defined estimands link trial objectives with design and analyses. This link is essential to ensure that trials yield results that inform the various stakeholder's decisions. The general, high-level definitions of primary objectives that have been commonly used in the past, such as "to evaluate the efficacy and safety of intervention X in patients with condition Y," are not sufficient because they leave room for ambiguity about exactly what is to be estimated (Phillips et al., 2016).

The need to quantify treatment effects from clinical trial results is obvious. The ICH E9(R1) Addendum elaborates on this point, defining treatment effects as how the outcome of an experimental treatment compares to what would have happened to the same patients under different treatment conditions, such as if they had not received the treatment or had received a different treatment. In other words, the goal is to establish the causal effects of well-defined interventions. Often, it is not feasible to assess the same patients under multiple treatments, as in a crossover design. However, randomization provides a statistical basis for establishing causal treatment effects even when each patient receives one treatment. This is a two-step process wherein randomization provides the causal link between patients and the treatment to which they are assigned, and it is assumed that patients follow the assigned treatment so that the causal relationship can be extended to the actual treatment taken (Ratitch et al., 2019a).

However, post-randomization events can compromise the causal links and protective effects of randomization against bias if those events are related to study treatment and outcome. Examples of relevant post-randomization events are premature discontinuation of randomized treatment or changing to or adding rescue treatment due to lack of efficacy or toxicity of the initially randomized intervention. Such events diminish or break the direct linkage between the randomized arm and taking the treatment and hence may lead to difficulties in assessing the causal effect of the initially randomized treatment by introducing confounding and bias.

Relying solely on randomization and ignoring the post-randomization events may lead to estimating the causal effect of being randomized to a treatment and not necessarily the causal effect of that treatment (Ratitch et al., 2019a). Therefore, whether certain post-randomization events can be ignored depends on whether inference is sought for the effects of the treatment as actually adhered to or in the effects of the treatment if it were adhered to.

If what has been termed a pure intention-to-treat (ITT) approach is followed in which all post-randomization changes of treatment are ignored, and all patients and all their outcomes are included regardless of treatment adherence, randomization is preserved, and missing data is minimized. However, this approach can confound treatment and outcome and break the link between randomization and the treatment taken and thus lead to difficulty in attributing causal effects to the initially randomized interventions. Therefore, an important aspect of the ICH E9(R1) Addendum and a topic further discussed in Chapter 3 are to define the treatment or treatment regimen of interest – the target of inference (Ratitch et al., 2019a).

The appropriate study design, data to be collected, and method of analysis can, and usually do, vary depending on the estimand. This issue highlights the necessity for updated guidance and the benefits from the recently proposed study development framework (Phillips et al., 2016) that begins with clearly defining objectives, which leads to clearly defined estimands, which in turn informs study design and analyses that are consistent with the objectives and estimands (Ratitch et al., 2019a).

With the new concepts of intercurrent events (ICEs) outlined in the ICH E9(R1) Addendum, older terminology may be obsolete. Carpenter et al. (2013) refer to treatment effects as actually adhered to as effectiveness or *de facto* effects, and treatment effects if patients adhered to the treatment as efficacy or *de jure* effects. *De jure* effects have also been termed counterfactual because assessments are based on the assumption that all patients were adherent while evidently in almost every realistic situation some patients will not be adherent.

In this context, *de facto* (effectiveness) estimands involve a combination of efficacy and adherence, whereas *de jure* estimands focus on efficacy only. The *de jure/de facto* and efficacy/effectiveness terminology have been conceptually useful constructs. However, as will become clear in subsequent chapters, the more nuanced ideas of ICEs laid out in the ICH E9(R1) Addendum render these terms obsolete. For example, multiple strategies for handling ICEs may be required in a single trial. Some of these strategies may be consistent with *de jure* (efficacy, counterfactual) ideas, while the other strategies used in the same trial may be consistent with *de facto* (effectiveness) ideas. Therefore, such a trial would involve a mix of *de jure* (efficacy, counterfactual) and *de facto* (effectiveness) strategies and labeling an estimand as *de jure* or *de facto* would be misleading. Therefore, throughout the remainder of this book, we try to avoid terminology such as *de jure* and *de facto* when describing estimands in favor of detailing the strategy or strategies of handing ICEs.

2.3 Considerations for Differing Stakeholders

In designing clinical trials, it is important to consider the decisions to be made by various stakeholders based on the evidence generated from the trial. These decisions should drive the objectives of the trial. For example, regulators decide whether treatments should be granted a marketing authorization. Payers decide whether and where a new drug belongs on a formulary list, whether the treatment is publicly funded and/or the appropriate reimbursement price (Mallinckrodt et al., 2017). Payer decisions typically focus on the price and benefit of the treatment relative to existing treatments. The costs and benefits are usually based on the intervention of prescribing the treatment rather than taking the intervention. Hence, regulators and payers make risk–benefit decisions, and each decision they make influences the treatment of many patients. However, they may have different perspectives because they make different decisions.

In contrast to the population-level decisions made by regulators and payers, prescribers make many decisions, with each decision influencing the treatment of one patient. Prescribers must also inform patients and/or care givers of what to expect from the prescribed medication. Patients/care givers each make one decision that influences the treatment of one patient (Mallinckrodt et al., 2017).

In general, those who make decisions about groups of patients are more interested in objectives and treatment effects based on the group of patients affected by the decision – even if these effects are an aggregated estimate from heterogeneous groups of patients. For example, a treatment effect aggregated from patients who do and do not adhere to the defined treatment regimen may be of interest to payers and regulators who make decisions on a population level. However, those making decisions about individual patients may not be interested in aggregated effects and will be more interested in patient-specific effects (Mallinckrodt et al., 2017). Moreover, perception of risk and/or benefit may also differ at the individual and population levels. An individual may be informed of "risk of cancer" and be deterred from using a treatment, even if that risk is low and does not preclude regulatory approval (Ratitch et al., 2019a).

2.4 Considerations for Differing Clinical Situations

In addition to differing perspectives for different stakeholders, other factors may influence choice of objectives. For example, perspectives on appropriate objectives may differ for inpatient versus outpatient studies or for disease-modifying drugs versus symptomatic treatments. Notably, objectives may depend on the stage of clinical development, therapeutic area, available treatment options, and so on. Phase I/II trials usually provide a proof of concept or identify doses for subsequent studies; that is, they primarily inform sponsor decisions. Phase III (confirmatory) studies typically serve a diverse audience and may need to address diverse objectives (Leuchs et al., 2015; Mallinckrodt et al., 2017).

This multifaceted nature of clinical trials is another important consideration in choosing objectives and estimands (Leuchs et al., 2015; Phillips et al., 2016; Mallinckrodt et al., 2017; Akacha et al., 2017). Often, multiple estimands will be needed to address the multiple objectives needed to inform the differing decisions of the various stakeholders. Even for a single stakeholder in a single trial, interest may lie in more than one objective and therefore require multiple estimands. For example, different estimands may be needed for safety than for efficacy. Proof of concept may be established for one estimand, but results from secondary estimands are needed to properly plan for phase III. Nevertheless, it is necessary to designate one objective and its corresponding estimand as primary, with other estimands being secondary or supportive. Regardless of the rank of the estimand, defining it clearly helps ensure that the relevant evidence is obtained and properly analyzed and interpreted (Ratitch et al., 2019a).

The following example described in Mallinckrodt et al. (2012) illustrates the benefits of assessing different estimands in the same trial. Drug A and Drug B (or dose A and dose B of a drug) have equal effectiveness, but drug A has superior efficacy and drug B has greater adherence. Dose/Drug A might be the best choice for patients with more severe illness because it has greater efficacy. Dose/Drug B might be best for patients with less severe illness and/or safety and tolerability concerns because it has greater adherence resulting from fewer side effects. A more nuanced understanding of efficacy and adherence could motivate additional investigation that could lead to more optimized patient outcomes. For example, subgroups of patients who especially benefit from or tolerate the high dose might be identified from the existing data or from a new trial (non-responders to low dose). Or, alternate dosing regimens that might improve the safety/tolerability of the high dose, such as titration, flexible, or split dosing (40 mg every 2 weeks rather than 80 mg every 4 weeks), could be investigated in subsequent trials.

2.5 Summary

A simple definition of estimand is what is to be estimated. Ambiguity in interpreting the results of medical research can arise when it is not clear what exactly is to be estimated in a statistical analysis. Estimands clarify what is to be estimated and link trial objectives with trial design and analysis. Clinical trials often involve multiple stakeholders who need to make different decisions. Therefore, multiple estimands are likely needed in any one trial. Moreover, perspectives on what is important to estimate may vary even for an individual stakeholder depending on the clinical context, such as stage of development and disease state.

Section II

Estimands

Section II begins with a chapter that details how clear estimands lay a foundation for specifying aspects of trial design, conduct, and analysis needed to overcome the issues caused by intercurrent events (ICEs). A study development process chart is presented as a tool to guide choice of objectives, estimands, study design, and analysis. The role of the intention-to-treat principle is discussed in this new framework. Emphasis is placed on defining the treatment or treatment regimen of interest and to define which post-randomization events (ICEs) are a deviation from the regimen of interest. Estimands for the special circumstances of safety endpoints, early-phase trials, non-inferiority trials, and quality-of-life endpoints are discussed. Finally, an overview of trial design issues is presented.

Chapter 4 details the five strategies for dealing with ICEs noted in the ICH E9(R1) Addendum: treatment policy strategy, composite strategy, hypothetical strategy, principal stratification strategy, and while-on-treatment strategy. Assumptions, implications, strengths, and limitations for each strategy are discussed. Chapter 5 provides three progressively more complex examples of choosing and specifying estimands in real-world clinical trial scenarios. Chapter 6 links classic clinical trial concepts with corresponding concepts and ideas from causal inference. Potential outcome nomenclature is used to precisely define estimands using mathematical notation, which can eliminate ambiguity in the definitions for those familiar with the notation. Chapter 7 revisits and summarizes key issues from Chapters 3 through 6 and presents a conceptual road map for dealing with ICEs.

3

Estimands and How to Define Them

3.1 Introduction

Clearly defined estimands lay a foundation for specifying aspects of trial design, conduct, and analysis needed to yield results that inform the stakeholder's decisions. General, high-level definitions of primary objectives such as "to evaluate the efficacy and safety of intervention X in patients with condition Y" are not adequate because they leave room for ambiguity about exactly what is to be estimated (Phillips et al., 2016). The focus of this chapter is on a systematic process for specifying objectives and estimands, and ensuring that the design, data, analyses, and sensitivity analyses are aligned with the objectives and estimands.

The authors credit James Bell and Johnathan Bartlett as co-authors through their contributions to the paper upon which this Chapter is based.

3.2 Study Development Process Chart

Ratitch et al. (2019a) built upon the previous study development process charts (Phillips et al., 2017) and the ICH E9(R1) Addendum (ICH, 2019) to propose an updated study development process chart in Figure 3.1. This more detailed process chart is needed to address the more nuanced approach in the ICH E9(R1) that expanded the discussion from missing data to the broader context of any post-randomization events that can confound assessment of treatment effects. The addendum introduced the term "intercurrent events" (ICEs) to define these post-randomization events.

The more detailed process chart begins with considering the objectives of the trial, which entails considering the decisions made by the various stakeholders from the trial results (Mallinckrodt et al., 2017). Subsequent steps include defining the primary estimand, followed by determining design, analysis, and sensitivity analyses.

The following text from Ratitch et al. (2019a) provides additional details on the specifics of choosing and defining estimands (Steps 1 and 2). Sections III and IV of this book cover Steps 3 and 4. Previous work stressed the need for

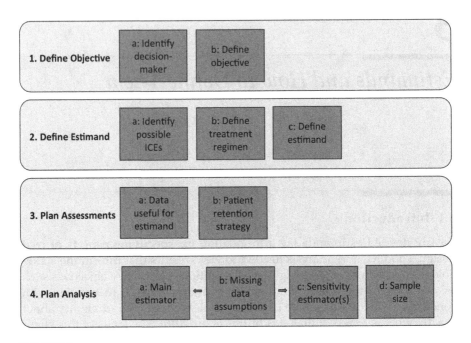

FIGURE 3.1
Study development process chart.

the study development process to be iterative to account for the interrelatedness of the items (Leuchs et al., 2015; Phillips et al., 2016). Although iteration may be necessary, with the more detailed process chart, the goal is to be as complete as possible during each step, thereby minimizing or avoiding the need for iteration (Ratitch et al., 2019a).

1a: Identify who will use the trial results and what decisions they will make from those results. This is an essential first step because the estimand must align with the decision-maker(s) needs. Any one trial may need to address the diverse needs of multiple stakeholders, leading to the need for multiple objectives and estimands.

1b: Define the general question of interest to the decision maker. This will typically include factors such as the interventions being compared, patient population, endpoint, and time scale.

2a: List all the ICEs that are plausible to occur in the trial, noting their likelihood of occurrence. Doing this early in the process avoids overlooking aspects of the intervention effect that could cause confounding or bias in analysis.

2b: Define in detail the treatment regimen under investigation. The definition should include whether interest is exclusively in the initially randomized treatment or in treatment regimens that include the initially randomized treatment. It is therefore also necessary to specify whether each ICE is part of the regimen under investigation or is a departure from the intended regimen.

2c: There are four components needed to define an estimand: the population, the endpoint, the summary measure, and how ICEs will be handled. Each component is important. However, we are familiar with the first three, and therefore focus here is on handling ICEs. Based on the treatment regimen defined in 2b, specify the strategy to handle each type of ICE. The ICEs that are part of the regimen are handled using the treatment policy strategy. For those ICEs that are deviations from the treatment regimen, use the clinical question to determine which ICEs are outcomes (e.g., dropout due to adverse events is considered treatment failure). The remaining ICEs are confounding factors that may be problematic for the assessment of the outcome and treatment regimen of interest. To complete the definition of the estimand, define the population, endpoint, and summary measure considering the classifications of ICEs and chosen strategies to handle them. A final step is to confirm that the estimand is aligned with the clinical objective and fits the stakeholder requirements.

The following subsections provide additional details on aspects of the study development process chart.

3.3 Process for Defining Estimands

3.3.1 Introduction

Properly defining an estimand requires clear specification of the population, the endpoint, the summary measure, and how ICEs are handled. The first three points are familiar and not discussed further here. Additional detail on handling ICEs is provided in subsequent sections, where it is assumed that the endpoint and summary statistic are clearly defined, along with a clear definition of the population that is also reflected in the trial's inclusion and exclusion criteria.

3.3.2 Identifying Intercurrent Events

ICH E9(R1) defines events that occur after randomization that alter the course of the randomized treatment as intercurrent (ICH, 2019). Therefore, ICEs are inherently connected to the definition of the treatment regimen. Some ICEs may not be a deviation from the treatment regimen of interest and therefore require no further consideration. Other ICEs are a deviation from the regimen of interest and must be dealt with in some manner, depending on the estimand of interest. For example, minor alterations in background therapy may be allowed as part of the treatment regimen of interest, but addition of or switching to a new medication would be a deviation from the regimen of interest. In other situations, adding or switching to a new medication may not be a deviation from the regimen of interest. As such, appropriate analyses are driven by the treatment regimen (target of inference) and the estimand of interest (Ratitch et al., 2019a).

ICH E9(R1) provides several examples of ICEs: use of an alternative treatment (e.g., a rescue medication, a medication prohibited by the protocol, or a subsequent line of therapy), discontinuation of treatment, treatment switching, or terminal events such as death (ICH, 2019). As noted in subsequent sections, how to properly handle an ICE may depend on the underlying reason. For example, it may be appropriate to handle the ICE of treatment discontinuation differently for those who discontinue for an adverse event versus those who discontinue for reasons not related to the treatments, such as relocating and no longer being able to travel to the investigative site.

While many ICEs are common across trials, some are specific to certain types of trials, such as incorrect food intake or vomiting in pharmacokinetic-based trials. The addendum also distinguishes between ICEs that make data observed after the ICE irrelevant versus those that lead to missing data. For example, if a patient died in a clinical trial where death is not the primary endpoint, the originally planned assessments will be incomplete, but further observation is neither possible nor relevant. This is an example of a change in patient state (Ratitch et al., 2019a). In contrast, nonadherence to the defined treatment regimen may mean that data after nonadherence is not useful in evaluating the primary estimand even though it is possible to observe the data (Ratitch et al., 2019a).

Both changes in state and deviations from the planned regimens are ICEs that must be anticipated and dealt with. We first consider deviations from the randomized treatment and then discuss changes of patient state. In discussing the former, we assume that it is theoretically possible to observe outcomes after the ICE on the same schedule as for patients who did not have an ICE and we consider how the ICE affects interpretation of these outcomes.

To ground the discussion on identifying ICEs and to distinguish those that are deviations from the treatment of interest, we consider the following example of patient profiles in Table 3.1 from Ratitch et al. (2019a). We assume that the trial has three post-randomization visits. Patients have to take randomized treatment through visit 3 – the primary endpoint. The profiles in Table 3.1 describe treatment courses where various events may cause deviations from the initial randomized treatment. The second column depicts treatments at each visit:

TABLE 3.1

Example Patient Profiles Depicting Various Intercurrent Events

	Visits			
	V1	**V2**	**V3**	**Treatment Received**
1.	X	X	X	Randomized treatment alone through the end of study (ideal adherence)
2.	X	O	O	Randomized treatment at V1; no treatment at V2, V3
3.	X	‡	‡	Randomized treatment at V1; randomized treatment + rescue at V2, V3
4.	X	+	+	Randomized treatment at V1; rescue alone at V2, V3
5.	X	‡	+	Randomized treatment at V1; randomized treatment + rescue at V2; rescue at V3
6.	X	‡	O	Randomized treatment at V1; randomized treatment + rescue at V2; no treatment at V3
7.	X	+	X	Randomized treatment temporarily interrupted for rescue at V2

Note: X = randomized treatment; O = no treatment (with the randomized treatment, experimental or placebo, discontinued, and no other treatment started); ‡ = randomized treatment with an addition of concomitant rescue; and + = rescue treatment without the randomized treatment.

X = randomized treatment

O = no treatment (with the randomized treatment, experimental or placebo, discontinued, and no other treatment started)

‡ = randomized treatment with an addition of concomitant rescue

+ = rescue treatment without the randomized treatment

Therefore, X indicates adherence to the randomized treatment at an assessment, with all other symbols indicating ICEs that deviate from the randomized treatment.

Profile #1 is a fully adherent patient with no deviations from the initially randomized treatment. All other profiles depict deviations from the initially randomized treatment. Profile #2 is a patient who receives randomized treatment for the first visit and no treatment for the last two visits. This profile is applicable only if it is ethical to leave patients untreated for part of the study, such as when no alternative treatments exist. Profile #3 adds concomitant rescue treatment to the initially randomized treatment. In Profile #4, patients discontinue randomized treatment and switch to the rescue treatment. Profiles 5, 6, and 7 are mixtures of the previous scenarios. In profile #7, randomized treatment is temporarily interrupted and substituted by rescue, and then randomized treatment is resumed (Ratitch et al., 2019a).

Randomization determines the initial treatment (X). The efficacy and toxicity (real or perceived) of X drive, at least to some extent, the adherence to X and subsequent changes in therapy. Deviations from the initial randomized treatment lie on the causal pathway between the randomized treatment and outcome at Visit 3. In all profiles but #1, the observed outcome at Visit 3 is influenced by the effect of the initially randomized treatment and the effect of some other therapy (or lack thereof). All these scenarios (and others) may occur in almost every clinical trial. Therefore, it is important to consider early in trial planning the possible ICEs (Ratitch et al., 2019a).

3.3.3 Defining the Treatment Regimen of Interest

Clearly defining the treatment regimen of interest (target of inference) is essential in providing the information needed by various stakeholders. The following ideas in this section are taken from Ratitch et al. (2019a) and explain why defining the treatment regimen of interest is important and how treatment regimens can be defined in practice.

When defining treatment regimens, descriptions should include doses or dose ranges for the initially randomized treatments and background therapies, allowed and prohibited concomitant medications, along with allowable rescue medications and their doses. One example of a treatment regimen is when patients initiate an experimental drug or placebo at randomization but discontinue treatment prior to the intended endpoint due to some contraindications (adverse reactions or lack of a response). The absence of alternative treatment options and/or the expected future course of illness may result in a plan not to treat such patients for the remainder of the study. In this case, the treatment regimen is the randomized treatment for a specified duration or until contraindication, with no other treatment taken after contraindication.

An example of a multitreatment regimen is when patients start with an initial randomized treatment but can switch to or add rescue medication if a minimal improvement is not achieved within some specified period. In this case, the treatment regimen is the experimental drug plus rescue versus placebo plus rescue, with specific conditions for when rescue should be initiated and for how long it should be taken.

The strict intent-to-treat (ITT) (treatment policy) approach is one specific example of a treatment regimen. Here, the treatment regimen is the initially randomized treatment with discontinuation of that treatment and/or use of any other treatment allowed at any time. It is therefore the act of randomization rather than the intervention that is causally linked to the outcome. This objective (target of inference) may be appealing in certain circumstances, such as for payers who want to link outcomes to financial costs. However, the target of inference is not a specific treatment regimen, nor is there guarantee that the regimens used in a trial will mirror general clinical practice where patients are neither randomized nor blinded to initial treatments (Mallinckrodt et al., 2017). Therefore, a detailed definition of the treatment regimens to be evaluated is essential because it lays out the expected course of treatment, including acceptable post-randomization flexibilities (see Section 3.8 for additional considerations regarding the ITT principle).

The target of inference (initial treatment or treatment regimen) is the "cause" to which the outcome is attributed. Some decision-makers' interest may be exclusively in the effects of the initially randomized treatment, while others may be interested in regimens that include the initially randomized treatment. Which treatment regimen is relevant for each objective should be assessed considering the clinical context, including natural evolution of the condition, the symptoms being treated, availability of alternative treatments, and their mechanism of action and their effectiveness.

Clarity in defining the treatment regimen of interest is an essential early step in the study development process. No universally best choice exists. In some cases, post-randomization modifications of the initial treatment are of interest. However, these modifications contribute to the observed effect, and therefore inferences and decisions apply to the entire regimen, not to its individual components. Therefore, the treatment regimen definition is the foundation for understanding which ICEs constitute a deviation from the regimen of interest and how to deal with them. Moreover, clearly defining the treatment regimen is also essential for determining an adequate sample size because the assumptions about the expected treatment effect size depend on the treatment.

Once the treatment regimen of interest is determined, possible ICEs can be categorized as either part of the treatment regimen or a deviation from it. For example, if the treatment regimen is defined as the randomized treatment plus background therapy and possible adjustments to it if specific criteria are met, then adjustments in background therapy are part of the treatment regimen, and this ICE is handled by the treatment policy strategy without additional considerations in the analyses.

3.3.4 Overview of Strategies for Handling Intercurrent Events

The ICH E9(R1) Addendum outlines five strategies to deal with ICEs: four strategies consider ICEs as a deviation from the treatment of interest: composite, hypothetical, principal stratification, and while-on-treatment. In the fifth strategy called the treatment policy strategy, ICEs are not a break from treatment (ICH, 2019).

For treatment profiles that deviate from the treatment regimen of interest, outcomes after such ICEs can no longer be causally attributed to that regimen alone. The ICH E9(R1) Addendum identifies four strategies for dealing with ICEs, which are a break from the planned treatment. The choice of which strategy or strategies to use depends on the estimand (Ratitch et al., 2019a). The following list is the definition of the strategies as provided in the ICH E9(R1) Addendum (ICH, 2019). Detailed explanations are provided in the subsequent sections.

Composite strategy: The occurrence of the ICE is a component of the variable; that is, the ICE is integrated with one or more other measures of clinical outcome as the variable of interest. ICEs that can be accounted for using a composite strategy are usually important clinical outcomes such as treatment discontinuation. For example, patients who discontinue early are classified as treatment failures while patients who adhere to treatment are classified as treatment successes or failures based on clinical responses.

While-on-treatment strategy: Response to treatment prior to the occurrence of the ICE is of interest. If a variable is measured repeatedly, its values up to the time of the ICE may comprise the endpoint, rather than the value at the same fixed time point for all patients. The timing of the ICE or duration of treatment up to the event is not relevant.

Hypothetical strategy: A hypothetical scenario of scientific interest is defined. The value to reflect that scientific question of interest is that which the variable would have taken in the hypothetical scenario. Some examples include the outcomes if patients had not stopped or switched treatment or outcomes if patients had taken rescue medication instead of discontinuing the trial.

Principal stratification: The patient population is defined as the principal stratum with a specific status in terms of ICE occurrence (i.e., the ICE does or does not occur) under specified treatment(s). For example, the stratum of patients who would have remained alive on either/all treatments.

How ICEs are handled depends on the estimand. The following examples provide a brief overview with additional details provided in the subsequent sections.

Scenario 1: Interest is in a treatment regimen that includes the initially randomized treatment with or without rescue medication. Early discontinuation is accounted for using a composite strategy in which patients who discontinue early are classified as treatment failures. Patients who remain in the study are classified as successes or failures based on predefined levels of improvement. Use of rescue medication is part of the treatment regimen of interest and is handled with the treatment policy strategy.

Scenario 2: Similar to scenario 1, except that the target of inference is the initially randomized treatment without rescue. Unlike scenario 1, here rescue is a break from the regimen of interest. Therefore, both early discontinuation and use of rescue are accounted for using a composite strategy wherein these patients are considered treatment failures.

Scenario 3: Similar to scenario 2, the target of inference is the initially randomized treatment. Both early discontinuation and rescue are accounted for using a hypothetical strategy because interest is in the effects of the initially randomized treatment if it were taken as directed. The aim is to estimate what outcomes would have been if patients had not stopped treatment or used rescue medication. Alternatively, a principal stratification strategy can be used wherein treatment effects are estimated from the principal stratum of patients who would not discontinue early or need rescue medication on either treatment.

Chapter 4 provides detailed discussion of each strategy, their implications, and when they may or may not be appropriate. Multiple strategies can be used for the same estimand, for example, with a composite strategy used for some ICEs and a hypothetical strategy used for others.

3.4 Defining the Estimand

The primary estimand should balance succinctness with clarity. The language should be understandable by clinicians and statisticians (NRC, 2010). The ICH E9(R1) lists the following aspects that together describe the estimand (ICH, 2019):

1. The population, that is, the patients targeted by the scientific question
2. The variable (or endpoint), to be obtained for each patient, that is required to address the scientific question
3. How to account for ICEs to reflect the scientific question of interest
4. The population-level summary for the variable that provides a basis for a comparison between treatment conditions

The elements in the estimand definition are interrelated and should be coherent. Choice for one element may influence the choice of another (Leuchs et al., 2015; Mallinckrodt et al., 2017). For example, the choice of endpoint influences the choice of the population-level summary that is appropriate for that variable (e.g., means versus proportions). The strategies chosen to account for ICEs need to align with the variable and population. These estimand elements must align with the specific treatment regimen under evaluation.

As seen in some of the examples in Chapter 5, achieving clarity in some estimands, even when defined as succinctly as possible, can be cumbersome. In these situations, it may be useful to summarize the estimand either before or after its full specification based on the four general categories outlined in Chapter 2. For example, estimands can be summarized as to whether inference is sought for the initially randomized treatment or regimens that involve the initially randomized treatment, the strategy or strategies used to handle ICEs, along with stating the endpoint and population-level summary statistic.

3.5 Special Considerations in Defining Estimands

3.5.1 Estimands for Safety Outcomes

Safety outcomes are sometimes a primary objective of a trial, but even when not primary, safety is always important to evaluate. The ICH E9(R1) estimand framework may be applied to safety outcomes although this has not been discussed in detail in the literature. As with outcomes that assess the benefit of a treatment, when assessing the risk of a treatment, it is important to define the population, outcome, summary measure, and treatment regimen of interest.

Issues may arise when integrating efficacy and safety as part of a benefit–risk assessment if the two derive from estimands with different treatment regimens. If efficacy was assessed based on a treatment regimen including rescue while safety was based on the initially randomized treatment only, then benefits and risks may not be directly comparable; the benefits of rescue are included without the drawbacks (Ratitch et al., 2019a).

3.5.2 Estimands for Early-Phase Trials

The estimand framework applies to all clinical trials because it is always necessary to define what is to be estimated. However, nuances and different considerations apply to certain types of trials. Estimands for early-phase trials may differ from the confirmatory settings focused on in ICH E9(R1) in several ways, most prominently regarding the stakeholder, and in some instances, the use of single-arm trials (Ratitch et al., 2019a).

Some early-phase trials are designed with regulatory stakeholders in mind (e.g., oncology trials) because they may form part of the submission package. These trials may have a primary estimand typical of confirmatory trials. However, most early-phase trials are designed primarily to inform the sponsor and the general scientific community about whether proof of concept has been established. Estimands for this purpose often focus on the initial treatment with less emphasis on treatment policies that may be the focus of confirmatory trials. Secondary estimands similar to those used in confirmatory trials can be evaluated in these early-phase trials to further inform development decisions and to plan for the subsequent trials.

A second key difference between early- and late-phase trials is that sometimes early-phase trials are single arm, that is, have no control arm. Estimands are usually defined as a comparison between treatments within the trial, but this is not possible in a single-arm trial. Instead, estimands need to focus on single-arm estimation and/or on comparison to a predefined (or historical) target. Although inferential difficulties in single-arm trials versus trials with concurrent control are well known, most of the considerations regarding dealing with ICEs are the same. In general, hypothetical and/or composite strategies may be most straightforward to compare the results from historical controls because their estimates are based on more clearly defined scenarios and thereby better suited to like-for-like comparisons, or if like-for-like comparisons are not possible, then it is easier to identify how conditions differ (Ratitch et al., 2019a).

3.5.3 Estimands for Scenarios When Treatment and Outcomes Do Not Occur Concurrently

Important considerations arise when treatment and outcome assessment do not occur concurrently. Examples of this are in oncology with chemotherapy, radiation, or other treatment lasting for a relatively short duration, and the primary assessment is based on overall survival time or more generally whenever short-term disease-modifying treatments have long-term outcomes (e.g., surgery). Clearly, assessments taken while on treatment will not capture the outcomes of interest.

Here, ICEs may be competing events, for instance, "other death" in an assessment of cardiovascular death. Definitions for censoring may include a mixture of ICEs. It is therefore critical to identify precisely what the trial is trying to estimate. The treatment regimen definition is also critical in what are often trials of long duration with multiple rescue treatment options. In oncology, interest may exist in questions regarding the sequencing of treatments or whether treatment may be stopped successfully after a certain period. Again, a clearly defined treatment regimen and estimand are essential (Ratitch et al., 2019a).

Historically, the oncology and other trials focusing on death or other major events as a primary outcome have used the treatment policy strategy that ignores ICEs in order to infer effects regardless of changes of treatment. However, differential use of rescue medication is likely and can attenuate the effects of the initially assigned treatments. Importantly, when the treatment policy strategy is used, inference about the initial treatment is not possible because the causal link between the initial treatment and outcome is broken. When focus is on the effects of the initial treatment, a hypothetical strategy may be useful. Typically, the hypothetical strategy requires more complex statistical methods that entail more assumptions than a treatment policy strategy. However, this trade-off can be justifiable if the simpler treatment policy strategy is not estimating the parameter of interest (Ratitch et al., 2019a).

3.5.4 Estimands for Quality-of-Life Evaluation in Trials with Many Deaths

In trials where many deaths are anticipated, such as oncology or heart failure studies, the evaluation of quality of life (QoL) is challenging. The QoL scales assess aspects of patient satisfaction and function while alive and do not have a designated value or category for death. Several potential questions of interest exist for these outcomes. A straightforward question, but one that can be complex to address, is QoL while alive. Duration of survival would be assessed separately.

An estimand for this objective could use a while-on-treatment strategy with respect to the ICE of death, with a summary measure that incorporates all available measurements up to death. Alternatively, treatment benefit at a specific time point may be of interest, for example, 1 year after the start of treatment, using a combined measure of QoL and survival. In this case, a composite strategy could be used for death as an ICE. However, this approach exemplifies the need for care in defining the composite endpoint. Assigning a numerical pseudo value for patients who die (e.g., a zero value as the lowest possible QoL score) may not be the best strategy. A zero for quality of life is not equivalent to death and could skew treatment differences and be difficult to interpret (Ratitch et al., 2019a).

Consider, for example, two cancer treatments. One prolongs survival but has severe side effects; the other has a smaller effect on survival but has few side effects. In a composite endpoint, the second treatment could appear to have lower QoL, driven by the greater number of zeros resulting from deaths, but while alive patients on that treatment had greater QoL due to lower side effects. This example brings into question the appropriateness of this composite strategy in this situation (Ratitch et al., 2019a).

An alternative would be to use a variable that represents the ranking with respect to the original QoL measurements if patients survive to the time point of interest and a worse rank (possibly depending on time of death) if patients die. Alternatively, a categorical variable can be used with category definitions based on the QoL scores and death defined as the worst category. Success of such an approach hinges on clear definitions of clinically meaningful categories. Other possible approaches include area under the curve (AUC) of the QoL scores over time until death, which would reflect the duration of survival or time-to-event endpoints such as time to QoL deterioration or death (whichever comes first) or QoL-weighted survival time.

Another objective may be to evaluate the QoL at a specific time point of interest for patients who survive to that time point. Here a principal stratification strategy could be used. Modeling is required to identify those patients who would survive to the time point of interest regardless of treatment assignment. This requires strong assumptions that survival is predictable from baseline characteristics and that all relevant covariates are measured (Ratitch et al., 2019a).

3.5.5 Estimands for Assessing Non-inferiority

Historically, certain analytic considerations were thought to apply differently to non-inferiority and superiority studies. For example, regulatory settings demand rigid control of type 1 error rate – false-positive findings. In the more common setting of assessing superiority, "conservative" analyses are preferred over liberal ones, where conservative means underestimating the advantage of the experimental drug over control or providing assurance that the effect is not overestimated, thereby ensuring the false-positive rate is no greater than the nominal level. However, in non-inferiority settings, underestimating the difference between treatments can inflate the false-positive rate because it is more likely to conclude the treatments do not meaningfully differ if there is a meaningful difference.

Consequently, sometimes estimands chosen for non-inferiority studies have differed from those that would have been chosen in an otherwise similar superiority study in order to avoid inflating false-positive findings. The study development process chart detailed in Section 3.2 and the higher-level process in the ICH E9(R1) make clear the limitations to that view on appropriate estimands and analyses for superiority versus non-inferiority studies. For example, changing the estimand based on analytic considerations has the process backward – the scientific question should drive the estimand and in turn the estimand drive the analyses. In the "traditional" approach, analytic considerations drive the choice of estimand, which in turn may alter the scientific question being addressed. Therefore, it is just as important for non-inferiority studies as for superiority studies to first define the estimand, based on the scientific question and then determine the analysis.

3.6 Trial Design and Conduct Considerations

3.6.1 Introduction

It is easy to appreciate in principle that trial design should be matched to the situation – that is, the primary estimand. However, given the new focus on ICEs, it is important to explicitly consider how the ICEs and the ways they are dealt with impact trial design. Specifically, a design should be chosen that enables estimation of the effect of interest with minimal assumptions (ICH, 2019). That is, the design should be driven by the objective and estimand and should enable appropriate data collection for primary and secondary estimands, including careful collection of data related to ICEs, concomitant medications, and discontinuation reasons that are part of the considered estimands. Ratitch et al. (2019a) detail the following ideas on trial design and data collection that enable appropriate evaluation of estimands.

3.6.2 Data Collection and Trial Conduct Considerations

Treatment policy strategies require the collection of as much data as possible even after discontinuation of randomized study treatments. Other strategies do not require postdiscontinuation data because it is either irrelevant (hypothetical) or unused (composite, principal strata, while on treatment).

If no estimands involving treatment policy are planned, it is not necessary to collect postdiscontinuation data. However, many strategies involve at least some element of treatment policy and there may be multiple estimands of interest in the trial.

For example, in early-phase trials focusing on the proof of concept, it may be useful to gather information about the outcomes for patients following study treatment discontinuation to inform future confirmatory trials. Even where post-ICE data are irrelevant for all reasonable estimands, off-treatment data can still provide information useful in sensitivity analyses: off-treatment data can inform imputation strategies and/or place limits on what is likely to have occurred had treatment continued. For example, patients' measurements following discontinuation of the initial treatment could be used as a conservative estimate of their performance had they continued treatment as planned. Therefore, standard practice should be to collect all post-ICE data, at least for the primary endpoint.

It is also important to collect information about the ICE themselves. For any estimand where distinction is drawn between types of ICEs, it is important that categorizations of type are accurate, predefined, and as objective as possible. Some distinctions are based on changes in treatment, and it is important to record patient medication usage and post-ICE outcomes. Other distinctions between ICEs may be based on the reason for study drug or study discontinuation; for instance, discontinuation due to lack of efficacy may constitute a different ICE than discontinuation due to a serious adverse event and discontinuation for reasons unrelated to treatment (such as logistical reasons) treated in yet another manner.

Where such distinctions are inherent to dealing with ICEs and therefore part of an estimand, these reasons should be defined a priori, and specific training should be provided to sites to ensure complete and accurate data. Uninformative reasons for discontinuation such as "withdrawal of informed consent," "patient decision," and "physician decision" should be minimized through trial design and conduct. To handle the occurrence of unforeseen types of ICEs, a blinded review of ICEs could occur concurrent with blinded review of protocol violations.

The treatment policy strategy also has an impact on trial conduct. The distinction between discontinuation of the initially randomized treatment and discontinuation from the study overall is fundamental to the implementation of this strategy. With a treatment policy estimand, patients should continue study follow-up as originally planned even if they are no longer taking the initially randomized treatment. This needs to be specified in the informed consent form and discussed with patients. If rescue medication is provided to study participants at no cost, it may motivate them to remain in the study. Obviously, supplying rescue medication increases trial cost. If alternative therapies are not provided as part of study procedures, patients will have to essentially consent to trial-related assessment burden while taking treatment that they could obtain in the same way at the same cost as outside of the study. Protocols may need to make provisions for a reduced assessment schedule to collect only the most essential information during post-treatment follow-up.

When implementing the principal stratification strategy, need may exist to collect a more extensive set of demographic, medical history, and baseline disease assessments to predict stratum membership with greater accuracy.

3.6.3 Trial Design Considerations

Some general principles are useful to consider when making specific design decisions for a trial. First, consider ICEs as belonging to one of the two categories: avoidable and not avoidable. The NRC expert panel report on missing data (NRC, 2010) emphasized the importance of limiting missing data. The principle should be extended to ICEs in general. For example, if patients need to discontinue study drug due to adverse events or lack of efficacy, so be it. However, discontinuations for reasons not related to study drug should be minimized by trial design and conduct (Hughes et al., 2012; Ratitch et al., 2019a).

A second principle is to minimize reliance on definitions and assumptions in statistical models and strive to collect data by design, for example, whenever possible use cross-over, run-in, or randomized withdrawal designs rather than relying on principal stratification. One should not rely on a hypothetical strategy if the data could be collected; nor should one rely on a "blanket" treatment policy strategy for all ICEs when it is possible to implement precisely defined treatment regimens that would be more relevant to clinical practice. One should also not rely on composite and while-on-treatment strategies simply because they limit missing data. It is not justifiable to change the question of interest for convenience (Fleming, 2011).

A third principle to consider is the trade-off between internal validity and external generalizability relative to the choices for the primary estimand. Randomization, blinding, placebo control, precisely scheduled follow-ups, additional tests/measures and interventions, rigid rules for changing doses, and so on are often essential in drawing inferences about treatment effects. However, these design elements may limit the generalizability of results to actual clinical practice because they are not part of clinical practice, and they can influence response rates, adherence to initially randomized medications, and the use of rescue medication (Akacha et al., 2017; Mallinckrodt et al., 2017). Therefore, in the highly controlled settings of blinding and placebo control, the generalizability to clinical practice of estimands where treatment changes are ignored or considered an outcome should be considered. In these situations, hypothetical strategies that consider what outcomes would have been observed had patients not changed treatment may be more relevant. The point here is not to argue for any one strategy because no universally best strategy exists. The key point is that the design should be chosen to match the objectives and estimands.

3.7 A Note on Missing Data

Throughout this chapter, discussion of missing data has been deliberately minimized because defining an estimand is based on ICEs rather than on missing data. Importantly, missing data is not an ICE; it is a consequence of ICEs. For example, early-study withdrawal is an ICE and a consequence of it is missing data. The estimand defines the property of the target population that is being measured, and consequently its definition is independent of missing data. All well-defined estimands are prone to missing data since measurements may be missing even in the absence of ICEs.

It is not appropriate to define an estimand whereby either missingness itself is an outcome or it is substituted by an arbitrary value because missingness is usually not of interest itself. Regardless of the estimand, some data may be missing, and all estimators require untestable assumptions to handle missing data. The main consideration in missing data is in the robustness of estimators to assumptions required to handle the missing data. The more missing data, the more sensitive estimates are to the assumptions. In this regard, hypothetical strategies are generally the most affected, because in addition to unobserved (missing) data, they also regard post-ICE data as irrelevant and hence effectively missing (Ratitch et al., 2019a).

Treatment policy strategies may also be sensitive to missing data because they require measurements post-ICE, which are often difficult to obtain because patients tend to withdraw from the study if they have an ICE that causes them to discontinue treatment. In contrast, estimators for estimands that employ while-on-treatment, principal strata, and composite strategies for ICEs are the most robust to missing data assumptions because they are defined in ways that mean post-ICE data are not needed (Ratitch et al., 2019a). Missing data is discussed in greater detail in Chapter 15.

Trials should aim to maximize adherence to protocol procedures including adherence to the initially assigned treatments. Maximizing adherence improves robustness of results by reducing the reliance of inferences on the untestable assumptions (NRC, 2010). Similarly, improving patient follow-up post-ICE improves the robustness of estimation of treatment policy estimands. However, complete follow-up is rarely possible, hence sensitivity to missing data assumptions should be checked with sensitivity analyses unless complete data are obtained. Sensitivity analyses are covered in Chapter 16.

3.8 A Note on the Intention to Treat Principle

The ICH E9 (ICH, 1998) guidance has often been cited when justifying what is now termed a treatment policy strategy for accounting for ICEs. The 1998 guidance emphasizes the ITT principle, which is defined as "the principle that asserts that the effect of a treatment policy can be best assessed by evaluating on the basis of the intention to treat a patient (i.e., the planned treatment regimen) rather than the actual treatment given. It has the consequence that patients allocated to a treatment group should be followed up, assessed and analyzed as members of that group irrespective of their compliance to the planned course of treatment."

Thus, ITT has two parts, the patients to include and the data for each patient to include (Permutt, 2015a; Mallinckrodt et al., 2017). Some deviations from one or both aspects of the intent-to-treat principles are routinely accepted (Permutt, 2015a). The guidance is clear on the need to include all randomized patients and that all data should be included, but it does not specifically address data after the ICEs that confound the effect of the randomized treatment, for example, rescue treatment.

Rescue therapy is specifically addressed in ICH E10 (2000). In referring to trials with rescue, E10 states: "In such cases, the need to change treatment becomes a study endpoint." Thus, according to E10, postrescue data need not be included in the primary analysis.

It is important to recognize that the ICH guidance was developed before the NRC report (NRC, 2010) that focused on the need for clarity on estimands. The E9 guidance refers to ITT as the best way to assess treatment policies/regimens. It does not address inference for the initially randomized interventions. Considerations of estimands today are often more nuanced than anticipated by the 1998 ICH guidance, and thus aspects of the 1998 guidance are no longer sufficient and the need for the ICH E9(R1) Addendum is clear.

A reasonable approach is to maintain the principles of ITT but employ a slight modification to address the need for greater specificity arising from the various strategies to account for ICEs (Mallinckrodt et al., 2017). In this modified approach to ITT, all randomized patients are again included, thereby maintaining consistency with the first tenant of ITT. The second tenant of ITT is modified to mean that all data – *relevant to the estimand* – are included. For example, when evaluating estimands where the target of inference is the initially randomized treatment, postrescue data is not relevant and is therefore not included in the analysis, thereby avoiding the confounding effects of rescue medication. (Note that principal stratification strategies would not include all randomized patients.)

O'Neill and Temple (2012) noted that treatment policy strategies for all ICEs may be a more common choice for the primary estimand in outcomes trials where the presence/absence of a major health event is the endpoint and/or the intervention is intended to modify the disease process. In these scenarios, it is often ethically necessary to provide rescue medication before observing the primary outcome (death or major health event). In contrast, symptomatic trials (symptom severity is the endpoint) often use a primary estimand where the inferential target is the initially randomized treatments. In these scenarios, the confounding from rescue medications can be avoided by excluding data after discontinuation of study medication/initiation of rescue from the primary analysis.

3.9 Summary

Post-randomization events that are related to treatment and outcome can break the causal link between the randomized treatments and the outcomes, thereby clouding inferences about the treatment effects. The study development process chart described in Section 3.2 is a tool to help deal with these difficulties. This process chart is a more detailed version that builds on fundamental concepts of earlier process charts. The new process chart follows the same general path of earlier iterations in first choosing objectives, then defining estimands, and then matching the design and analyses (primary and sensitivity) to the objectives, estimands, and design. The new process chart provides additional details that are helpful in addressing how ICEs should be handled and is therefore consistent with the new ICH E9(R1) Addendum guidance.

Clinical trials may have many stakeholders, and objectives may vary for different stakeholders because they make different decisions. With clear objectives, it is then possible to define the treatment regimen of interest and to define which post-randomization events (ICEs) are deviations from the regimen of interest. Although many specific estimands are conceptually possible, four general scenarios account for most of the options. These four scenarios are defined by two domains: whether inference is sought for the initially randomized treatment or regimens that involve the initially randomized treatment and whether the inferences pertain to the effects (of the treatment or treatment regimen) if all patients adhered to it or as it was actually adhere to. The appropriate study design, data to be collected, and method of analysis can, and usually do, vary across these four general scenarios.

Specifying estimands entails defining the population, the endpoint, the summary statistic, and how ICEs are to be handled. Estimands for some special circumstances are discussed. Finally, an overview of trial design issues is presented.

4

Strategies for Dealing with Intercurrent Events

4.1 Introduction

The research regarding missing data and methods to account for it is extensive. However, the ICH E9(R1) Addendum moves the discussion beyond missing data to the broader context of intercurrent events (ICEs). Shifting to this broader context is essential in defining and interpreting estimands to address the varying needs of different stakeholders. The following subsections provide definitions of and considerations for use of the five strategies for dealing with ICEs that are outlined in the addendum. These ideas closely follow Ratitch et al. (2019a).

The authors credit James Bell and Johnathan Bartlett as co-authors through their contributions to the paper upon which this Chapter is based.

4.2 Treatment Policy Strategy

In the treatment policy strategy, ICEs are not a break from the treatment regimen of interest and therefore can be ignored. Adopting a treatment policy strategy for all ICEs is what has become known as a "pure ITT approach." In this approach, groups are compared without making allowances for lack of adherence, adding or switching to rescue medication, changes in background medication, and so on. Although pure ITT seems simple, it entails complexities that impact study conduct, sample size, cost, and interpretation of results. This is also true, to a lesser extent, when applied to only some of the ICEs.

The treatment policy strategy can only be applied if it is possible, at least in principle, to observe the outcome as planned (ICH, 2019). This strategy is not applicable when the ICE is death. Since death is at least always a possibility in a trial, the treatment policy strategy is technically undefinable unless death is the outcome. However, this strategy may still be relevant when few deaths are expected and survival is not expected to be affected by study treatment.

When used for all ICEs, the treatment policy strategy broadens the treatment regimens under evaluation because whatever treatment is taken or not taken, whenever it is taken or not taken, is part of the treatment regimens of interest. Such loosely defined comparisons may not be meaningful because the target of inference is not clearly defined, and the treatment regimens used in blinded and randomized clinical trials may not correspond and generalize to real-world practice.

The treatment policy strategy is perhaps more useful if applied to specific types of ICEs, for example, profile #2 from Table 3.1, premature discontinuation of the initially randomized treatment with no subsequent alternative treatment. The treatment regimen under evaluation is the initially randomized treatment – with possibly imperfect compliance – but no other therapeutic interventions. This is arguably the least controversial and most beneficial application of the treatment policy strategy, when it is ethical not to have rescue treatment, or when no alternative therapies exist.

A more common and complex situation is the addition of or switch to another therapy. In placebo-controlled trials, patients can cross-over from placebo to the experimental treatment, or placebo- and drug-treated patients can switch to a treatment approved for the indication. In general, these changes of treatment are for ethical reasons and may not reflect the clinical question of interest. If the rescue and/or treatment switching is of interest, then the treatment regimen under evaluation is a multitherapy, dynamic regimen that evolves based on treatment effects and with patient and clinician preferences.

One motivation for using the treatment policy strategy for the ICE of addition of or switch to rescue is that changes in and additions of treatment happen in clinical practice, and the prescribing information in the label should reflect drug effects that will be seen in clinical practice. For the treatment policy strategy to be useful, the decisions for whether, when, and what additional therapies may be initiated must either be predefined or, if based on the investigator and patient preferences, representative of real-world practice. This is a noteworthy consideration in blinded and placebo-controlled trials because these factors are never present in real-world practice. It may not be reasonable to expect patients and clinicians to have preferences and make decisions in a clinical trial as they would in general clinical practice because they do not know what drug is being taken, which in placebo-controlled trials means, they do not know if the patient is taking an active drug or not (Mallinckrodt et al., 2017). Another consideration is that clinical practice may change over time such that the regimen tested in a clinical trial, although once relevant, no longer reflects current practice.

In multiarm comparative trials, the same strategy should be applied for an ICE in all treatment arms. However, that does not mean the strategy will have the same effect on each arm and thereby not influence treatment contrasts. It is likely that more patients take rescue in a less efficacious treatment arm, such as placebo. Ironically, the lack of efficacy in the placebo arm will make the experimental treatment look relatively less efficacious due to the greater use of rescue in the placebo arm (Mallinckrodt et al., 2017).

Even where both arms are equally likely to require rescue, providing a shared treatment between arms for a proportion of the treatment time may reduce effect sizes by reducing the difference between the treatment regimens being compared. For instance, in oncology trials, it is standard practice to move patients onto standard of care treatment following tumor progression. Treatment policy strategies confound the effects of the initially randomized treatments on overall survival because both arms share post-progression treatment, thereby potentially reducing the magnitude of the difference between arms.

A reduced difference between treatment regimens makes it easier to demonstrate non-inferiority or equivalence between arms. In such cases, it may be difficult to make decisions about the comparative merits of treatment regimens without separately considering the proportions of patients who used rescue. Non-inferiority margins based on historical data would also need to account for the treatment policy strategy, which may be difficult to do accurately.

The treatment policy strategy may be appropriate for handling the ICE of rescue medication when rescue has minimal impact on the endpoint being investigated. For example, while receiving a therapy to prevent asthma flares, the primary endpoint, patients may need occasional inhaler use, which would not influence the number of flares. The treatment policy strategy may also be useful when there is a synergetic effect of the initial treatment with rescue and rescue alone does not provide satisfactory relief. An example is treatment of migraines where patients use multistep treatment strategies, starting with one medication and adding another, which may be needed for some portion of a patient's migraine attacks.

Sample size calculations should be based on the anticipated effect of the treatment regimen and may differ from that which would be seen for the initial treatment alone. If initiation of rescue does not follow strict predefined rules, greater variance in the outcomes may also be anticipated due to variable use of rescue. Because use of rescue tends to decrease differences between treatment arms, it will also tend to increase sample size – and therefore trial cost – for a given power. Ironically, use of rescue, an ethical necessity, when using the treatment policy strategy can result in exposing more patients to inferior or ineffective treatment (Mallinckrodt et al., 2017).

4.3 Composite Strategy

In composite strategies, the occurrence of ICEs is combined with an outcome measure to form a single composite variable. This strategy was mentioned in the ICH E10 (2000, Section 2.1.5.2.2), regarding trials with rescue: "In such cases, the need to change treatment becomes a study endpoint." More generally, when an ICE is a break from the defined treatment regimen, a bad outcome is assigned for that patient.

A common example of this strategy is where the study endpoint is defined as a binary (responder/non-responder, treatment success/failure) variable based on the degree of change in clinical measures or patient-reported outcomes. Patients with an ICE, such as premature discontinuation of study treatment or need for rescue, are classified as non-responders/failures regardless of the degree of change in the outcome measure. The observed outcomes determine response/success status for patients who did not have a relevant ICE.

Combining the occurrence of ICEs with numerical outcome measures is less straightforward than for binary outcomes. One possibility is to assign an unfavorable value of the outcome measure for patients with an ICE and then include those patient's assigned value along with observed values from patients who did not have an ICE in the population-level summary of the numerical variable. The choice of this surrogate value and the corresponding summary measure is important so that the distribution of the composite variable is not unduly skewed or its variance inflated. For example, in an endpoint that typically ranges from zero to ten, with low values representing bad outcomes, assigning values of 0 for patients with relevant ICEs will likely have minimal effect on the distribution because the assigned values are similar to actually observed values. In contrast, if the typical range of an outcome is 60–80 on a 0–100 scale, assigning a value of 0 will inflate the variance and skew the distribution. It may also be difficult to clinically justify that value. It may be necessary to use either robust or nonparametric statistical methods to deal with the skewed distribution. Alternatively, the numeric variable can be transformed into ranks, and patients with an ICE assigned a poor rank. This approach may be easier to justify clinically.

If all patients who have an ICE are assigned the same outcome, it is assumed that the outcome is equally bad for all patients with the ICE and that partial effects prior to the event are irrelevant. Alternatively, rankings for ICEs can be based on timing and severity of adverse events, degree of insufficient response leading to initiation of rescue, and so on.

Composite strategies are appropriate for ICEs that are clearly bad outcomes, such as death or serious adverse events. However, composite strategies are more problematic for ICEs that are not clearly a bad outcome, for example, early termination of study drug. If this resulted from a serious adverse event, classifying patients as non-responders (treatment failure) makes sense. However, if the patient discontinued because they were overly burdened by trial participation, calling them a treatment failure does not make sense.

The composite strategy can impact sample size requirements in two ways. First, if the underlying numerical measure is dichotomized to accommodate a composite outcome or if a less sensitive summary measure is used, sensitivity is lost, and a larger sample sizes are needed. Second, composite approaches favor the treatment arms with fewer ICEs, leading to need for smaller or greater sample sizes, depending on which arm has fewer ICEs. However, rates of ICEs vary even under similar trial designs. Hence, the direction and magnitude of these effects can be difficult to anticipate.

4.4 Hypothetical Strategy

Randomized clinical trials are controlled experiments to evaluate causal treatment effects under specific conditions. For various practical and ethical reasons, it is not always possible to implement, or for patients to follow, the desired conditions. Hypothetical strategies are used to address relevant questions when the desired experimental conditions cannot be ensured for all patients.

Covariate adjustment is a common example of a strategy to compensate for imperfect experimental conditions. It is rare to observe perfect balance between arms, but covariates are often fit to estimate the treatment effect under the hypothetical scenario that the arms had equal proportions of patients for categorical covariates and equal means for continuous covariates.

Other hypothetical strategies can remove the effects of treatment changes that would otherwise confound estimation of the treatment effect of interest (just as covariate adjustment removes confounding caused by imbalance).

Although the number of hypothetical scenarios that might be considered is extensive, most of the practical applications fall into a few categories:

- Outcomes expected if patients did not have ICEs that constituted a break from the treatment or treatment regimen of interest – that is, if all patients had adhered to the treatment regimen
- Outcomes expected if patients had followed a specific treatment regimen, typically involving a change in treatments, that either was not available in the trial or was not followed by some patients if available

The descriptions of hypothetical scenarios should contain the details of what is hypothesized to occur and not to occur (ICH, 2019). For example, in a placebo-controlled clinical trial in Type 2 diabetes, it is not ethical to withhold rescue treatment because that would prolong insufficient blood glucose control. However, the use of rescue medication masks or enhances (i.e., confounds) the effects of the initial medication. If stakeholder questions relate to the initial medication, hypothetical strategies can be useful to estimate the treatment effect expected if patients (contrary to the fact) had not used rescue. It is always important to describe the hypothetical conditions reflecting the question of interest.

The hypothetical scenario assuming that all patients were adherent is sometimes criticized because it is counter to the fact that some patients were not adherent (Permutt, 2015a). However, the question of treatment effects if patients adhered is important (Mallinckrodt et al., 2014, 2017). Patients are always advised to take their medication as directed. To develop appropriate directions, the effect of the drug if taken as directed must be evaluated. Put in another way, when patients and doctors are trying to decide whether a medication should be taken, it is necessary to know what is expected to happen when the medication is taken. A hypothetical strategy is needed because it is not ethical to ensure full adherence by design. We therefore abstract from the observed experiment to the question of interest via appropriate analytical methods.

Estimands based on if all patients adhered are of interest in many situations, including in earlier phase or mechanistic trials where interest is in a pharmacological treatment effect that *can* be achieved with ideal adherence (Mallinckrodt, 2013; Mallinckrodt et al., 2017). Another example where a hypothetical strategy to assess an estimand based on if all patients adhered may be useful is in non-inferiority or equivalence trials that must be designed to conform with the prescribing guidelines of the active control. It is therefore of interest to evaluate the efficacy of both the control and experimental treatment "if taken as directed." Potential differences in tolerability of the two treatments can be assessed separately. For later-phase trials assessing biomarkers rather than outcomes, such as HbA1c in diabetes or the FEV1/FVC measures of lung function in various respiratory trials, the effect of taking treatment as directed on the biomarker is of interest, particularly since further real-world outcome evidence is usually required for, or after, regulatory approval.

For those who make decisions about individual patients, knowing what happens if patients adhere may be more relevant than the effects in a mix of adherent and nonadherent patients. In contrast, for those who make decisions about groups of patients, the counterfactual nature of estimands if all patients adhere may not be relevant (Mallinckrodt and Lipkovich, 2017; Mallinckrodt et al., 2017).

Even when interest is primarily in estimands based on what happened as actually adhered to, secondary interest may be in effects if all patients had adhered (Mallinckrodt and Lipkovich, 2017; Mallinckrodt et al., 2017). For example, in oncology trials, the current standard treatment policy strategy for assessing the overall survival (OS) includes the effects of changing treatment following progression, which is ethically required. Adopting a hypothetical strategy focusing on the effect of the initial treatments on OS allows the benefit (or lack thereof) of the initial medication to be clearer.

Most of the discussion on hypothetical strategies has been in the context of assessing drug benefit. However, estimands for assessing drug risk are also important, for example, a drug that increases blood pressure. Some patients may become hypertensive and discontinue study medication and/or take medication to treat the high blood pressure, with subsequent return to normal blood pressure. Estimands that assess the drug's effects as actually adhered to reflect the patients' return to normal, thereby suggesting no change at the planned endpoint of the trial and no adverse effect of the drug. Estimands using the hypothetical strategy of assessing effects if patients had remained adherent to the initial treatment would not reflect a return to normal and would reflect increases at endpoint because had the patients been adherent, they would likely have continued to be hypertensive (Mallinckrodt et al., 2017).

Like composite or treatment policy strategies, hypothetical strategies can be limited to a portion of ICEs, for example, not all premature discontinuations of the initial randomized treatment occur for treatment-related reasons. It may be reasonable to consider what would happen if patients who discontinued for reasons not related to treatment had continued with treatment. Patients who discontinued for reasons related to treatment could be handled using a composite strategy, for example. Hypothetical strategies are especially relevant in long-term studies where attrition due to study fatigue and assessment burden may increase with time.

A scenario where the hypothetical strategy is likely not meaningful is when the ICE is death when assessing quality of life. Estimating hypothetical quality of life in patients who are dead can lead to irrelevant abstractions. For trials with few expected deaths or when death is not related to treatment or to the disease under investigation, the interpretation remains acceptable (Permutt, 2015a). A combination strategy could address interpretational issues around death by having most ICEs dealt with via a hypothetical strategy with death considered a bad outcome via a composite strategy.

Hypothetical strategies typically require unverifiable assumptions and statistical modeling techniques to predict outcomes under the hypothetical conditions due to either missing or irrelevant data. Such unverifiable assumptions are not unique to hypothetical strategies and occur in any analysis with missing or irrelevant data (such as postrescue outcomes when assessing effects of the initial medication).

4.5 Principal Stratification Strategy

Another alternative is principal stratification wherein treatment effects are assessed in the stratum of patients who would have a specific status with respect to the ICE (ICE would or would not occur) under one or more treatments in the study. It is not appropriate to adjust for post-randomization variables in the same way as adjusting for baseline characteristics because of post-randomization selection bias. Principal stratification is a framework in which patients' membership in the principal strata is defined in a way that is not affected by treatment assignment, and consequently the principal strata can be used similar to pretreatment stratification variables (Frangakis and Rubin, 2002; (see Chapters 6, 12, and 24 for additional details on principal stratification).

With two randomized treatments, four principal strata can be defined with respect to a specific ICE: patients who would experience an event on both treatments (A), patients who would not experience an event on either treatment (B), and patients who would experience the event on one treatment but not the other (C and D). Comparing treatments *within* each stratum (where occurrence of an ICE does not depend on treatment) yields a causal treatment effect (ICH, 2019).

The ICH E9(R1) suggests that the target population of interest may be the principal stratum in which an ICE would not occur, that is stratum A (ICH, 2019). An example where this strategy may be useful is when interested in the effect of treatment in patients who would adhere to either treatment, experimental and control. With respect to rescue medication, a principal stratum could be patients who would not require rescue medication on either treatment. Another example is evaluating a causal effect of vaccine on viral load in "patients who would be infected," that is, who would become infected regardless of randomization to placebo or vaccine (Ratitch et al., 2019a). In this example, it is not known at baseline/randomization who will become infected and the subgroup who became infected could be a nonrandom subset. Therefore, principal stratification can foster unbiased estimation of treatment effects, whereas the subgroup analysis based on the observed subgroup membership under the patients' actual treatment could be biased.

Another application of principal stratification is when dealing with death where death is not the endpoint of interest, for example, when evaluating the effect of treatment on quality of life where a non-negligible proportion of patients die before the time point of interest (Rubin, 2000). In this case, a meaningful treatment effect, referred to as survivor average causal effect (SACE), is well defined in the principal stratum of patients who would have survived to a specific time point on either treatment. Another potential example for symptomatic treatments would be to focus on the principal stratum of patients who cannot tolerate or otherwise adhere to placebo – in other words, those who are clearly in need of pharmacologic treatment.

In a parallel group design, it is not possible to directly observe which stratum patients belong to because we observe what happens only on the one treatment to which patients were randomized. The principal stratum of patients who would not discontinue from treatment regardless of which treatment they are randomized to is not the same as a subgroup of patients who completed randomized treatment in a parallel-group trial (ICH, 2019). Completer analyses in parallel group trials compare treatments in different populations. Using the previously defined strata, a completer analysis compares strata B and C versus strata B and D. This may represent healthier patients who were able to complete on placebo versus less healthy patients who needed active treatment to complete. Unbiased comparisons require treatments to be compared in the same population (ICH, 2019).

When stratum membership cannot be observed, it must be predicted from models based on prerandomization covariates. As with any model, the predictions are imperfect (ICH, 2019). The original set of randomized patients is not fully preserved in principal stratification because the analyzed samples are subsets, but the modeling attempts to maintain causality by ensuring balance.

The relevance of principal stratification for informing future clinical practice should be considered. Prescribing physicians must be able to determine whether their patients match the profiles of the clinical trial population before they prescribe the treatment. Similarly, where the target population is not definable upfront, the evidence of benefit in the target population needs to be supported by both the probability of being within the target population and the/risk if the patient does not fall within it.

One situation where it is particularly important to consider the clinical relevance of principal stratification is in dealing with the ICE of early discontinuation in placebo-controlled trials, especially the principal stratum of patients that would adhere to both placebo and the experimental drug. The stratum of patients who would adhere to placebo may be a less severely ill subset and not ideal candidates for treatment with the experimental drug. Similarly, when comparing an experimental drug whose aim is improved tolerability to a standard of care, it will not be useful to focus on the stratum of patients that would adhere to both drugs.

4.6 While-on-Treatment Strategy

In the while-on-treatment strategy to account for ICEs, response to treatment prior to the occurrence of relevant ICEs is the primary focus. For repeated measures, values up to the time of the ICE may be most relevant, rather than the value at a landmark time point. For example, patients with a terminal illness may discontinue a symptomatic treatment because they die, yet the success of the treatment before death is still relevant (ICH, 2019) (e.g., for palliative treatments).

While-on-treatment strategies are generally appropriate only when duration of treatment is not important, which may occur either because duration is truly not important clinically or because the rate of an event or outcome is constant. One common example is the rate of an adverse event per unit time, which assumes a constant hazard. If this assumption is appropriate, the adverse event may be analyzed as number of events per X years of patient exposure. Other scenarios where while-on-treatment strategies are appropriate include time-to-event endpoints where a constant hazard can be assumed or when a constant rate of change (e.g., slope) can be assumed (Mallinckrodt et al., 2019).

Other while-on-treatment approaches include estimands with an endpoint of change to last on-treatment observation regardless of when that occurred. Every patient has a last on-treatment observation. Therefore, the while-on-treatment strategy yields no missing data due to ICEs.

Treatment effects can also be based on summing or averaging values prior to the occurrence of relevant ICEs.

4.7 Assumptions Behind the Strategies for Dealing with Intercurrent Events

4.7.1 General Assumptions

All strategies require assumptions around missing data whenever relevant data are missing, or available data are irrelevant. Although this affects hypothetical strategies the most, it also includes composite strategies where relevant data are missing but no ICE occurred, and for the treatment policy strategy when, as is usually the case, it is not possible to achieve 100% follow-up.

Historically, emphasis was placed on ensuring that assumptions were conservative – not biasing in favor of an experimental treatment but rather against it. While this is often relevant, the assumptions also need to be biologically plausible for the trial conclusions to be meaningful and useful to decision makers (Ratitch et al., 2019a). The loss incurred by an incorrect decision needs to be gauged as well.

The discussion of assumptions underlying strategies for dealing with ICEs in this section is rather informal. More rigorous and mathematical treatment can be found in later chapters and references therein. For example, assumptions required for identifiability of treatment effects within principal strata for specific estimation strategies are considered in detail in Chapter 24; assumptions underlying certain hypothetical strategies are discussed in Chapter 22.

4.7.2 Treatment Policy Strategy Assumptions

In treatment regimens that include more than just the initially randomized treatments, the actual regimens taken may vary across the sample. For instance, patients adhere to the initially randomized treatment, or progress onto another of potentially many allowed treatments. An assumption is therefore required that patients follow regimens in the trial in a similar, or at least relevant way, compared to those in the "real world" of clinical practice.

4.7.3 Composite Strategy Assumptions

As noted previously, composite strategies may in various manners incorporate adherence or tolerance with efficacy or functional outcomes. This entails an implicit assumption that adherence decisions in the trial are like those in clinical practice (Mallinckrodt and Lipkovich, 2017; Mallinckrodt et al., 2017). This consideration is especially important in trials with placebo, randomization, and/or blinding because these factors are never present in clinical practice (Mallinckrodt and Lipkovich, 2017; Mallinckrodt et al., 2017).

In placebo-controlled trials, the rates of discontinuation for active treatments may be higher than when the same treatments are tested in blinded trials not including placebo. If the measures used to engender adherence in the clinical trial are not feasible in clinical practice, the trial could yield biased estimates of effectiveness (*de facto* estimands) relative to the conditions under which the drug would be used (Mallinckrodt and Lipkovich, 2017; Mallinckrodt et al., 2017).

4.7.4 Hypothetical Strategy Assumptions

Hypothetical strategies yield an estimate for the detailed scenario only, typically based around following the specified treatment regimen. This scenario must be well specified, including what patients are assumed to do, and assumed not to do. The scenario must also address a scientifically/medically relevant question (ICH, 2019). Hypothetical strategies sometimes entail statistical models with assumptions that cannot be fully backed by the data (from the same or historical trial). Therefore, sensitivity analyses are important in assessing robustness to departure from the assumed conditions.

4.7.5 Principal Stratification Assumptions

Principal stratum strategies rely on covariates to determine (typically imperfectly) to which stratum each patient belongs when in fact the strata are not observable (ICH, 2019). To ensure preservation of causality and unbiasedness, it is assumed that the probabilities of stratum memberships can be correctly estimated by including all relevant factors in the model (Ratitch et al., 2019a). In this regard, principal stratification and hypothetical strategies are alike in that they both require predicting something that was not observed.

If principal stratification is used to account for the ICE of study drug discontinuation or use of rescue, then the implicit assumption that adherence and rescue decisions in the trial are like those in clinical practice again applies because the adherence decisions influence stratum membership.

4.7.6 While-on-Treatment Strategy Assumptions

While-on-treatment, estimands typically require that the timing of ICEs is not relevant. In practice, this may mean that while-on-treatment strategies are useful for outcomes that are summarized by a rate and/or hazard that is constant over time (Ratitch et al., 2019a). It is therefore important to assess this constancy assumption (e.g., linearity of change over time).

4.8 Risk–Benefit Implications

When evaluating the risk–benefit profile of a treatment, it is important to keep in mind the estimand and the strategies for ICEs. If assessing benefit using the treatment policy or composite strategy that results in outcome measures that combine efficacy with adherence and/or safety, then comparisons of benefit versus risk may lead to double counting the risks if these estimands are weighted against a separate evaluation of adherence or tolerance. To avoid this double counting, the hypothetical strategy corresponding to all patients adhering to the initial treatment could be used for benefit and separately consider risk, for example, the proportion of patients who discontinue the regimen. Alternatively, a treatment policy strategy could be used without a separate consideration of discontinuation data because adherence is already included in the assessment of effectiveness (Ratitch et al., 2019a).

Principal stratification assesses benefit in a subset of all randomized patients. This raises the question of what safety data are most relevant – in all randomized patients or in the subset in whom benefit was assessed.

4.9 Summary

Five strategies for dealing with ICEs are proposed in the ICH E9(R1) Addendum:

- **Treatment policy strategy**: In the past often based on a pure definition of ITT, wherein no ICEs are a break from treatment regimen of interest; that is, all ICEs are ignored. A more nuanced usage is now advocated where this strategy is used only for selected ICEs with a well-defined clinically relevant treatment regimen of interest.

- **Composite strategy**: The ICE itself provides all the necessary information about the treatment outcome for those patients who experience ICEs. Data after the ICEs provide no additional information. ICEs that can be accounted for using a composite strategy are usually important clinical outcomes such as treatment discontinuation due to lack of efficacy.

- **Hypothetical strategy**: Several possibilities exist. Some examples include what would have happened if the ICE had not occurred and what would have happened under some alternative regimen, such as if rescue had not been taken. Here, the ICEs are a confounding factor.

- **While-on-treatment strategy**: Outcomes up to the ICE provide all necessary information about the treatment outcome.

- **Principal stratification**: The patient population is defined as those patients with a prespecified status of ICE occurrence under one or more study treatments (irrespective of their actual treatment assignment).

Different types of ICEs may be handled with different strategies for different estimands. For instance, rescue therapy can be an outcome, while discontinuation of study treatment is a confounding factor. Or, use of rescue and early discontinuation could both be confounding factors, and in a third scenario, both could be considered outcomes.

The appropriateness of these strategies depends on the clinical context and decision-making objectives. These strategies also have implications for trial design and conduct, especially regarding what data needs to be collected. Implications of the various strategies on risk–benefit assessments should also be considered. For example, principal stratification assesses benefit in a subset of all randomized patients. This raises the question of what safety data are most relevant – in all randomized patients or in the subset in whom benefit was assessed. Moreover, composite strategies often lead to an outcome that combines benefit and adherence – a measure of risk. If so, care is needed in risk–benefit assessments to avoid double counting risks that would occur when using a composite endpoint as the measure of benefit.

5

Examples from Actual Clinical Trials in Choosing and Specifying Estimands

5.1 Introduction

This chapter utilizes the concepts outlined in previous chapters to illustrate examples of choosing and specifying estimands in real-life settings. The examples closely follow the ideas and content of a recent paper (Ratitch et al., 2019b).

Three progressively more complex settings are illustrated. The first example is a proof-of-concept (PoC) trial in a major depressive disorder (MDD). The key stakeholder is the sponsor who must make a business decision about whether the drug has sufficient potential benefit to continue development. The second example is a confirmatory trial in asthma. The third example is a confirmatory trial in rheumatoid arthritis. These are more complex examples that involve a variety of intercurrent events (ICEs) that are handled using multiple strategies. The estimands chosen for these examples are not the only acceptable choices for the respective scenarios. The intent is to illustrate the process and key concepts rather than focus on the specific choices. Therefore, readers should refer to the study development process chart illustrated in Figure 3.1 and the brief review below.

Recall that an estimand describes the quantity to be estimated to address a specific study objective. The four elements that together comprise the estimand definition are the population, the variable (endpoint), the population-level summary (parameter), and how to account for ICEs to reflect the scientific question of interest.

The first three elements (population, endpoint, summary measure) have typically been specified in study protocols, albeit not as part of an estimand definition. How to handle ICEs is a new requirement. Recall that ICEs are post-randomization events that represent changes in treatment during and/or affect our ability to measure or interpret outcomes and their causal links with the treatment regimen(s) of interests. Examples of ICEs include premature discontinuation of the randomized treatment, initiation of or switching to rescue therapy. ICEs can undermine randomization, confound the effects of the randomized treatment regimen, and thereby compromise the evaluation of causal effects of the randomized treatments.

The ICH E9(R1) stipulates that the estimand definition must specify how the types of ICEs that are likely to occur will be handled so that causal effects estimated in the presence of those events can be clearly interpreted. The ICH E9(R1) suggested the following five strategies that were detailed in Chapter 4 that can be used to handle ICEs:

- Treatment policy
- Composite
- While on treatment
- Hypothetical
- Principal stratification

The choice of the strategy is driven by the treatment regimen that is targeted for evaluation and depends on the clinical context. When the treatment policy strategy is used for an ICE that marks the start of a new treatment, the new treatment is part of the regimen being evaluated, in addition to the randomized treatment. This new treatment may be a switch to or addition of a new treatment or a period of no treatment (e.g., when the originally randomized treatment is discontinued, and no alternative therapy is administered). The other four strategies are used when ICEs mark the start of treatment changes that are not part of the treatment regimen being evaluated. The confounding these treatment changes introduce, if not accounted for, makes it difficult or impossible to derive useful conclusions about causal effects of the treatment or regimen for which inferences are sought.

Estimands should be defined early in the trial design process: after identifying decision maker(s) and their objectives and before determining the assessment schedule and choosing analysis methods and estimating sample size. The need to define estimands before choosing estimators is one of the major points emphasized in the ICH E9(R1) and is necessary to avoid situations where habitual choices of analytical methods lead to implicit estimand definitions, resulting in estimation of causal effects that are not aligned with study objectives and needs of decision makers. In other words, if the estimand is not clearly identified first, the choice of analysis will implicitly define the estimand, and the objective, when, of course, it should be the other way around with the objective and estimand determining the appropriate analysis.

The estimand definition is not expected to provide details of a statistical analyses, beyond a succinct specification of the population-level summary measure. After defining the estimand(s), the estimators (statistical analysis methods) are chosen so that they are aligned with the estimand. The strategies specified in the estimand to handle ICEs determine which data are useful for the estimand and, therefore, influence when and how data should be collected. Some data useful for the estimand may be unobserved (missing) and strategies for handling the missing data should be specified as part of the estimator's methodology, along with sensitivity analyses to assess robustness of conclusions to the assumptions inherent to the chosen method of accounting for missing data.

The authors credit Nita Goel and Johnathan Bartlett as co-authors through their contributions to the paper upon which this Chapter is based.

5.2 Example 1: A Proof of Concept Trial in Major Depressive Disorder

5.2.1 Background

MDD is a common psychiatric condition with a lifetime incidence of approximately 15% (Kessler et al., 2005). The disorder ranges from mild to severe and is associated with significant potential morbidity and mortality, contributing to suicide and adverse impact on concomitant medical illnesses, interpersonal relationships, and work. The objectives of treatment are to reduce or resolve signs and symptoms of the disease, restore psychosocial and occupational function, and reduce the likelihood of relapse or recurrence (Primary Care Clinical Practice Guideline, 2010). Guidelines support pharmacological therapy for the treatment of depression in addition to psychotherapy. Antidepressant medications include selective serotonin reuptake inhibitors (SSRIs), serotonin/norepinephrine reuptake inhibitors (SNRIs), atypical antidepressants, serotonin-dopamine activity modulators (SDAMs), tricyclic antidepressants (TCAs), and monoamine oxidase inhibitors (MAOIs) (Practice Guideline for The Treatment of Patients With Major Depressive Disorder, 2010).

Antidepressants in established classes (e.g., SSRIs, SNRIs) typically demonstrate initial benefits after 3–4 weeks of treatment. The current standard for short-term efficacy trials in MDD is randomized, double-blind, placebo-controlled, parallel designs of 6–8 weeks duration (FDA Guidelines, 2018). Expectations are that patients of all severities will be evaluated, but that clinical trials in patients with treatment-resistant depression (usually defined as having failed two or more pharmacologic therapies) will be conducted separately.

Clinical trials in MDD pose many difficulties. High rates of placebo response and premature discontinuation of the randomized treatment are key factors in MDD trials that limit the ability to distinguish an effective drug from placebo (Potter, Mallinckrodt, and Detke, 2014). This high rate of false-negative results is an especially important consideration in PoC trials. Another consideration for PoC trials generally (not specific to MDD) is that at this early stage of development, the optimum dose, dosing regimen, and/or formulation may not be known. Suboptimal dosing in the PoC trial could reduce treatment effects. However, knowledge gained from that trial could result in improved dosing and improved outcomes in subsequent trials.

5.2.2 Trial Description

The example trial was a randomized, double-blind, parallel-group Phase II, PoC trial in MDD (Goldstein et al., 2002). The treatment duration was 8 weeks in an adult outpatient population. The primary efficacy outcome was the Hamilton Depression Rating Scale (HAMD) 17-item total score (Hamilton, 1960) at the end of the double-blind treatment period. Efficacy assessments were planned at baseline, and each post-baseline visit at weeks one-eight.

At this early stage of development in this indication, treatment regimens involving other drugs are not relevant. Moreover, with many drugs already on the market for MDD, new drugs are likely to be used in difficult-to-treat patients that have not responded to or been intolerant of other drugs, making assessments including rescue therapies less relevant. Therefore, no rescue therapy was planned to be made available in the trial.

5.2.3 Primary Estimand

We now follow the study development process discussed in Chapter 3 for the primary objective of evaluating efficacy.

1a. *Identify decision maker*: The key decision maker is the sponsor.

1b. *Define objective*: The general objective is to determine whether the experimental drug has benefit in treating patients with MDD.

2a. *Identify possible ICEs*: The ICE of greatest concern is early discontinuation of treatment. Although lack of adherence due to intermittent noncompliance is a concern, it is difficult to evaluate. Pill counts are not considered a reliable indicator. Plasma drug concentrations, if available, provide some indication of compliance. For the purposes of this example, noncompliance other than early discontinuation is ignored. Actual treatment regimens that may occur in this trial in either a planned manner (per the study treatment discontinuation guidelines mentioned above) or unplanned are summarized in Table 5.1. Scenario 1 is adherence to the randomized treatment through 8 weeks of the double-blind period, without any ICEs. Treatment changes such as switching to no treatment or to standard of care, represented by Scenarios 2 and 3, respectively, are ICEs that may occur at any time during the 8-week double-blind period. Concomitant use of other antidepressant medications is prohibited. Deaths are not expected.

TABLE 5.1

Anticipated Treatment Regimens for a PoC Trial in MDD

Scenario	Treatment Regimen Over 8-Week Period
1	Z
2	Z → O
3	Z → P(i)

Note: Z = randomized treatment; O = no treatment; P = post-discontinuation of randomized treatment; (i) = nonstudy, standard of care treatment for MDD. Scenarios 2 and 3 result in missing data for this trial.

2b. *Define treatment regimen under evaluation*: The treatment regimen to address the objective stated in (1b) is the initially randomized treatment taken as directed for the planned duration of 8 weeks.

2c. *Define estimand*: The estimand is defined as follows, specifying the four elements as outlined in ICH E9(R1):

   a. The treatment effect is to be estimated for the population of adult patients with MDD as defined by the protocol inclusion/exclusion criteria.

   b. Efficacy is to be measured using change from baseline to week eight of the double-blind study period in HAMD 17-item total score.

   c. All ICEs leading to changes in treatment, such as premature discontinuation of the randomized treatment with or without a switch to alternative therapies, will be handled by a hypothetical strategy to estimate what the treatment effect would have been at the designated time point if all patients adhered to the initially randomized treatment through that time point.

   d. The difference between treatment groups in mean changes from baseline to week eight of the double-blind study period in HAMD 17-item total score.

3a. *Data useful for estimand*: The data necessary for this estimand is the observed data while adhering to the initial randomized treatment. Observations after discontinuation of the randomized treatment, regardless of initiation of subsequent therapies, are not useful for evaluating this estimand. Therefore, the data after such ICEs will not be collected for the purposes of this estimand. For patients with ICEs, week eight data will not be available, and the corresponding outcomes will be estimated consistent with the hypothetical scenario stated in the estimand definition (see more details in Step (4a) below).

Although data post-ICEs are not required for the primary estimand, it may be useful to collect these data for estimation of supportive estimands that can inform subsequent trials.

3b. *Patient retention strategy*: Retention strategies can focus on trial conduct features to minimize missing data. These features go beyond our current scope and have been discussed elsewhere (NRC, 2010).

4a. *Main estimator*: An estimator aligned with the estimand is a likelihood-based repeated measures approach, such as mixed model for repeated measures (see Chapter 18). The model will be fit to all available data collected from all randomized patients at scheduled assessments during adherence to randomized treatment, that is, through week eight or the latest time point prior to an ICE. Alternative estimators include inverse probability weighting or multiple imputation based on the MAR assumption followed by either analysis of variance (ANOVA) at the primary time point or repeated measures analysis of the longitudinal data (see Chapters 19 and 20).

4b. *Missing data assumption*: In this trial, some intermittently missing data may be expected due to patients occasionally missing a study visit while continuing with the randomized treatment. Moreover, data post-ICEs are not useful for the primary estimand even if collected and will be treated as missing. For both types of missing data, the primary analysis model assumes that patients with missing data would have efficacy outcomes like those of similar patients in their treatment group who continue their randomized treatment through the time point at which data are missing. This type of assumption is referred to as missing at random (MAR) (Little and Rubin 2002; Verbeke and Molenberghs, 2000) (see Chapter 15 for additional details on missing data assumptions).

4c. *Sensitivity estimators*: Sensitivity analyses need to be conducted to assess the robustness of conclusions from the primary analysis to the assumption that missing data arise from an MAR mechanism. This assumption can be stress tested via a delta adjustment tipping-point sensitivity analysis (see Chapter 23 for additional details).

4d. Sample size required for the primary estimand using the primary estimator assumes the effect size if all patients adhered. Subsequent trials may need to allow for additional margins for sensitivity analyses or be based on different estimands.

Although the level of specificity above is needed to fully understand what is to be estimated and therefore **must** be included in the study protocol, following this full specification it may be easier to appreciate the general concept by also stating a concise summary of the essential elements such as to assess the effects of the initially randomized treatments using a hypothetical strategy to assess what would be expected if ICEs had not occurred, based on mean changes from baseline on the HAMD17 in all randomized patients.

5.3 Example 2: A Confirmatory Trial in Asthma

5.3.1 Background

Asthma is a heterogeneous chronic inflammatory respiratory disease that impacts over 300 million people worldwide. Characterized by symptoms of wheezing, shortness of breath, chest tightness, and/or cough and accompanied by variable expiratory airflow limitation, asthma ranges from mild-to-severe disease associated with compromised quality of life and reduced survival (Global Asthma Network, 2014). Goals of asthma management include achieving symptom control, maintaining normal levels of activity, and minimizing future exacerbations to avoid long-term morbidity and mortality (Global Initiative for Asthma, 2018). Early treatment increases the likelihood of improved asthma control and less additional asthma medication use.

In addition to addressing modifiable risk factors and nonpharmacologic approaches, patients often step up pharmacologic therapy with increasing doses and potency of inhaled corticosteroids (ICSs), leukotriene receptor antagonists (LTRAs), theophylline, and long-acting beta$_2$-agonists (LABAs) based on continued symptomatology, receiving short-acting beta$_2$-agonists (SABAs) as needed. Those with continued symptoms may receive additional therapy with oral corticosteroids (OCSs) and/or anti-immunoglobulin E (IgE) or anti-interleukin 5 (IL-5) (Global Asthma Network, 2018).

In clinical trials of new add-on treatments for patients with severe asthma uncontrolled with high-dosage ICSs and LABAs, a placebo-controlled add-on design (standard therapy plus experimental drug versus standard therapy plus placebo) with a provision for a short-term rescue medication is the preferred approach (Busse et al., 2008). Marketing approval of new medicines is typically based on the primary efficacy measure of clinically significant asthma exacerbations rate (EMA, 2015). Clinically significant exacerbations of asthma are usually defined as a requirement for systemic corticosteroids or an increase of the maintenance dose of oral corticosteroids for at least 3 days and/or a need for an emergency visit, hospitalization, or death due to asthma. A clinical trial of approximately 1-year duration is required by regulators (EMA, 2015). Pulmonary function is typically assessed either as a co-primary or a key secondary endpoint and is predominantly measured as change from baseline in the pre-bronchodilator forced expiratory volume in one second (FEV_1) at a landmark time point during double-blind treatment.

In asthma trials of add-on therapies, the standard asthma controller background therapy consists of an ICS/LABA formulation. The prestudy dosage and regimen are continued throughout the study treatment period. Other allowed asthma controllers (e.g., long-acting muscarinic antagonists [LAMAs], LTRAs, and OCSs) taken at least 30 days prior to enrollment are usually allowed during the study, but typically, prior exposure to biologic therapies would not be permitted or require washout. SABAs via a metered dose device are also typically permitted as needed for worsening asthma symptoms, that is, for occasional short-term rescue use. However, a regularly scheduled or prophylactic (e.g., prior to planned exercise) use of SABAs in absence of asthma symptoms is typically discouraged. Other changes to treatment are also typically discouraged or disallowed during the study treatment period, for example, changes to the patient's background controller regimen and use of LABAs as a reliever (e.g., Symbicort® Maintenance and Reliever Treatment). Asthma exacerbations are normally treated with oral or other systemic corticosteroids according to the standard practice, and the protocols typically outline the exacerbation treatment guidelines.

5.3.2 Trial Description

For illustration, consider the SIROCCO trial of benralizumab (Bleecker et al., 2016). In this section, we summarize the main features of this trial and the primary estimand used as the basis for marketing approval of benralizumab as we infer it from the study publication (Bleecker et al., 2016) and publicly available regulatory marketing application review documents (FDA Application Number: 761070Orig1s000, Clinical Review(s), 2017a; Food and Drug Administration, Center for Drug Evaluation and Research. Application Number: 761070Orig1s000, Statistical Review(s), 2017b; CHMP. Fasenra Assessment Report, 2017). We also discuss another estimand that may be of interest for supportive purposes.

The SIROCCO trial was a randomized, placebo-controlled Phase 3 study of patients with severe asthma uncontrolled with high-dosage ICSs and LABAs. Patients were randomly assigned to one of the three treatment groups: subcutaneous benralizumab 30 mg either every 4 weeks (Q4W) or every 8 weeks (Q8W), or matching placebo. Patients received study drug injections at clinical centers every 4 weeks for the duration of 48 weeks. Planned assessment times included the randomization visit (Week 0) and visits at 4-week intervals during the treatment period (Weeks 4, 8, 12, …, 48). The primary endpoint was the number of asthma exacerbations evaluated over 48 weeks. A key secondary endpoint was the change from baseline to

Week 48 in the prebronchodilator FEV_1. The study sample was drawn from an enriched population of patients with blood eosinophil counts of at least 300 cells per microliter at baseline. The objective was to assess the effect of benralizumab as an add-on treatment; therefore, patients continued taking their background asthma controller treatments with a stable prestudy dosage and regimen during the study treatment period. The allowed rescue therapy, discouraged/disallowed medications, and management of exacerbation events were similar to the typical setting of an add-on treatment trial for severe asthma patients described above.

5.3.3 Primary Estimand

We now follow the steps of the clinical trial design process chart to reconstruct the primary estimand in the SIROCCO study.

1a. *Identify decision maker*: The primary decision makers were the regulatory agencies.

1b. *Define objective*: The primary objective was to assess the effect of benralizumab compared to placebo as an add-on maintenance treatment in patients with severe asthma.

2a. *Identify possible ICEs*: Treatment regimens that may have been anticipated in the SIROCCO trial, occurring in either a planned or unplanned manner, are summarized in Table 5.2. All scenarios except scenario 1 represent ICEs that occur at the time point when the treatment changes from the randomized treatment (Z) to either the prestudy background therapy only (O) or a different treatment regimen. Using the notation in Table 5.2, C(i) is protocol-allowed rescue therapy for short-term management of worsening asthma symptoms, whereas treatment changes (ii) and (iii) were discouraged or not allowed. Any of the treatment changes could occur at any point in the trial and their handling in the primary estimand did not depend on timing.

Typical study treatment completion rates in similar studies range between 80% and 85%, with higher rates observed in more recent confirmatory studies. The treatment completion rates in the SIROCCO trial were 89%, 87%, and 90% for the three treatment groups, respectively (Bleecker et al., 2016). Approximately 8% of patients took disallowed concomitant medications (Food and Drug Administration, Application Number: 761070Orig1s000, Clinical Review(s), 2017). Withdrawals and important protocol deviations were balanced across treatment groups (CHMP. Fasenra Assessment Report, 2017).

2b. *Define treatment regimen under evaluation*: The treatment regimen under evaluation was the randomized treatment taken for up to 48 weeks as add-on to background ICS/LABA, including

TABLE 5.2

Anticipated Treatment Regimens in the SCIROCCO
Trial of Severe Asthma

Scenario	Treatment Regimen Over 48 Weeks
1	Z
2	Z → O
3	Z → C(i)
4	Z → P(i)
5	Z → P(ii)
6	Z → C(ii)
7	Z → P(iii)

Note: Z = randomized treatment as add-on to prestudy
ICS/LABA regimen; O = background prestudy ICS/
LABA regimen only; C = concomitantly with the randomized treatment; P = postdiscontinuation of randomized treatment; (i) = SABAs for worsening
asthma symptoms as rescue and protocol-specified
treatment for exacerbation events; (ii) = changes to
the patient's background controller regimen, regular
or prophylactic use of SABAs, treatment with short-
acting anticholinergics or with oral or injectable
corticosteroids outside of managing an asthma
exacerbation event, use of LABAs as a reliever;
(iii) = nonstudy alternative treatment for asthma.
In addition to the above treatments, exacerbation
events were managed per the study-defined treatment
protocol.

protocol-defined rescue therapy and treatment for exacerbation
events, as well as other asthma treatments per investigator and
patient decision. In other words, the regimen of interest was the
originally randomized treatment and all treatment modifications
that may arise during the trial without prespecification. That is,
Scenarios 1–7 in Table 5.2 were all scenarios consistent with the
regimen of interest.

2c. *Define estimand*: The primary and secondary estimands were
not explicitly defined for the SIROCCO trial. Based on the reported
methods and results (Bleecker et al., 2016), we infer the following as
the primary estimand:

a. The treatment effect in adult and adolescent patients with severe
asthma uncontrolled with high-dosage ICSs and LABAs as
defined by the protocol inclusion/exclusion criteria who had
blood eosinophil counts at entry of at least 300 cells per microliter.

b. The primary endpoint was the number of asthma exacerbations
during the 48-week double-blind study period.

c. All types of ICEs, including use of SABAs for worsening asthma symptoms as rescue, treatment of exacerbation events as specified in the study protocol, premature discontinuation of the randomized treatment, and any modifications of asthma treatment including those that were discouraged/disallowed by study protocol but might have occurred per the investigator and patient decision, were handled using the treatment policy strategy, that is, included in the treatment regimen under evaluation.

d. The number and rate of asthma exacerbation events were calculated for each randomized treatment group based on the data collected over the 48-week post-randomization period, and each of the experimental treatment groups was compared to the placebo group using the event rate ratio.

The ICEs corresponding to the use of SABAs for worsening asthma symptoms as rescue and treatment of exacerbation events are protocol-defined treatments that are part of the standard of care recommended for ongoing disease management in this patient population. Based on their mechanism of action, and on considerable prior clinical experience, these therapies are not expected to produce lasting disease-modifying effects. Apparently, a treatment policy strategy was used for all ICEs. That is, no ICEs were a break from the treatment regimen under evaluation. Using the treatment policy strategy for all ICEs seems disconnected with the trial design in that the protocol explicitly stipulated medications that were discouraged or not allowed, but the analytic approach disregarded this fact. These ICEs may have been thought to likely occur in general clinical practice in a small percentage of this patient population. Therefore, the treatment effect estimated in presence of these ICEs was not expected to be significantly confounded and was considered clinically relevant for the evaluation of benralizumab.

3a. *Data useful for estimand*: Because the treatment policy strategy was used for all ICEs, usable data for this estimand were the exacerbation-related data over the 48-week post-randomization period regardless of adherence to the randomized treatment. Patients who switched to an alternative asthma treatment after they discontinued from the randomized treatment were expected to complete the remaining study visits. Patients who had post-randomization treatment changes that were discouraged or disallowed by the protocol were not withdrawn and were continued to be followed as planned.

3b. *Patient retention strategy*: The completion rate in the SIROCCO trial was greater than similar historical trials, which may be indicative of a general trend in recent years where regulatory and sponsor emphasis on improved patient retention bears positive results.

However, specific details of patient retention tactics used in the SIROCCO trial were not available in the documents we reviewed.

4a. *Main estimator*: The SIROCCO trial used a common approach for analysis of rates of exacerbation events that was based on a negative binomial model for the recurrent event data. The logarithm of patient follow-up time was used as an offset variable in the model to adjust for different follow-up times (Bleecker et al., 2016). The response variable was the number of exacerbation events for each patient during the 48-week double-blind treatment period. Typical models for such an analysis would include treatment group, baseline covariates, and randomization stratification factors.

4b. *Missing data assumption*: Although using the treatment policy strategy for all ICEs may reduce the amount of missing data compared to other strategies, missing data nevertheless occur due to patient withdrawal. Missing data is a confounding factor that needs to be dealt with in the analysis. The primary analysis with the negative binomial model assumed that missing data arose from an MAR mechanism (see Chapter 15); that is, patients who withdrew from the study were expected to have a similar exacerbation rate postwithdrawal to the exacerbation rate of patients in the same treatment group who remained in the study and who had similar values of baseline characteristics and other covariates included in the model.

4c. *Sensitivity estimators*: Sensitivity analysis focusing on the assumptions about missing data can be done by varying assumptions about the rates of exacerbations after early-study withdrawal. For example, it could be assumed that patients with missing data from the experimental arms had a greater exacerbation rate postwithdrawal than those who withdrew from the placebo arm (Keene et al., 2014). A range of such assumptions can be considered to find a tipping point at which the experimental group is no longer significantly different from control (see Chapter 16).

Another option for assessing sensitivity to missing data is imputation based on the model estimated from patients who discontinued the randomized treatment but remained on study and therefore have postdiscontinuation data. This option is most aligned with the primary estimand but would not be practical in the SIROCCO trial because only about 1% of patients were in this reference group, thereby leaving the imputation model poorly informed.

4d. *Sample size*: Some ICEs marked treatment changes that may lower the risk of exacerbation events. When data after such ICEs are used for the estimation of the overall treatment effect, the estimated treatment difference may be attenuated, and this should be considered in the sample size calculations at the trial design stage.

5.3.4 Supportive Estimand

The primary estimand provided an assessment of the treatment policy involving the experimental treatment, which in this case may be considered as estimating a lower limit of the experimental drug's benefit. A supportive estimand can provide an estimate of an upper limit of the drug's benefit, thus enabling a decision maker to evaluate a spectrum of evidence. This can be achieved by estimating a treatment effect during adherence to randomized treatment, allowing only for the protocol-defined rescue therapy and treatment of exacerbation events. The benefit of the randomized treatment while taken as directed can subsequently be interpreted considering separate analyses of safety, tolerability, and adherence. This supportive estimand would also be consistent with the stipulations of allowed and not allowed medications in the protocol, that is, using the treatment policy strategy only for ICEs of the type "$Z \to C(i)$" in Table 5.2. We now follow the trial design process chart to define this supportive estimand.

1a. *Identify decision maker*: Decision makers are again regulatory agencies.

1b. *Define objective*: The objective is to evaluate efficacy of the experimental drug compared to placebo as an add-on maintenance treatment in patients with severe asthma while the treatment is taken as directed.

2a. *Identify possible ICEs*: Potential ICEs are the same as discussed for the primary estimand and listed in Table 5.2.

2b. *Define treatment regimen under evaluation*: The treatment regimen under evaluation is the randomized treatment taken as directed for up to 48 weeks, including use of SABAs for worsening asthma symptoms as rescue and treatment of exacerbation events as specified in the protocol.

2c. *Define estimand*: The population and summary measure remain the same as for the primary estimand, while the endpoint and handling of ICEs are different from the primary estimand, using a combination of the treatment policy and while-on-treatment strategies:

Endpoint: Efficacy is to be measured using the primary endpoint of the number of asthma exacerbations for each patient while he or she receives the randomized treatment as directed, possibly with occasional uses of SABAs and management of exacerbation events as permitted by the protocol, up to 48 weeks of the double-blind study period.

To account for ICEs:

C1. For patients who require SABAs as rescue treatment for worsening asthma symptoms or treatment of exacerbation events as specified in the protocol, outcomes observed during the period of these additional treatments while continuing the randomized treatment as directed provide nonconfounded evidence for the effect of the treatment regimen under evaluation. Therefore, the treatment policy strategy is used with respect to these types of ICEs.

C2. For patients who initiate any other changes to the treatment regimen, including any asthma treatments or changes to the background controller therapy, which are discouraged/disallowed by the protocol, or premature discontinuation of the randomized treatment, outcomes after the ICEs that mark the start of such treatment changes are irrelevant for the evaluation of the treatment regimen of interest. A while-on-treatment strategy will be used with respect to these ICEs.

Note that this estimand does not target the treatment effect over the full intended study period of 48 weeks for all patients. It assesses the treatment effect during the period of adherence only, regardless of the duration.

3a. *Data useful for estimand*: Data useful for this estimand are the observations collected while patients adhere to the randomized treatment and take additional treatments only for rescue or management of exacerbation events as permitted by the protocol. All data after discontinuation of the randomized treatment and data after ICEs that mark the start of discouraged or disallowed treatment are not useful for this estimand and are excluded from analyses. (Additional information on how many patients used discouraged or disallowed treatments in each treatment arm would be helpful in interpreting results of this estimand.)

Steps 3b. *patient retention strategy*. No changes compared to what was discussed in the context of the primary estimand.

4a. *Main estimator*: The main estimator for analysis of exacerbation rates is the same as for the primary estimand, the only difference being the data included in the analysis as discussed in (3a).

4b. *Missing data assumption*: For the endpoint of the number of exacerbation events in the context of an estimand with while-on-treatment strategy, the only patients with missing data is those who were randomized but had no post-baseline data. The estimator assumes the missing completely at random (MCAR) mechanism (see Chapter 15). That is, the subset of patients with no post-baseline follow-up would have similar while-on-treatment outcomes to patients with post-baseline data in their treatment group who have similar baseline covariates. Although the MCAR assumption is generally questionable, missing data in this context are expected to be rare.

4c. *Sensitivity estimators*: Similar stress-testing sensitivity analyses as for the primary estimand could be employed.

4d. *Sample size:* Similar considerations apply as for the primary estimand.

5.4 Example 3: A Confirmatory Trial in Rheumatoid Arthritis

5.4.1 Background

Rheumatoid arthritis is a systemic inflammatory autoimmune disease impacting approximately 0.5%–1% of the population. Severity ranges from mild to severe and can include progressive joint destruction, compromised quality of life, and reduced survival (Smolen and Steiner, 2003; Colmegna, Ohata, and Menard, 2012). Remission is the optimal treatment goal because it is correlated with the prevention of structural damage and maintenance of function (Felson et al., 2011; Smolen et al., 2016). Early and aggressive treatment increases the likelihood of disease control (Klareskog, Catrina, and Paget, 2009); however, remission rates are low despite significant advances in the treatment over the past two decades. Patients often receive one or more conventional disease-modifying antirheumatic drugs (cDMARDs) with methotrexate (MTX) considered the gold standard. Often, cDMARDs are used in combination with low-dose oral or intraarticular corticosteroids and nonsteroidal anti-inflammatory drugs (NSAIDs). Moreover, biological agents that antagonize critical inflammatory mediators, T cells or B cells, are used with or without concomitant cDMARDs.

We consider a confirmatory trial of a biologic agent for MTX inadequate responders (MTX-IRs). In such studies, the primary and key secondary endpoints are measures of symptom improvement and physical function measured at or after 12 weeks of treatment. Per the FDA draft guidance for marketing approval of drug products for treatment of RA (FDA Guidance for Industry, 2013), demonstration of efficacy should include clinical response and physical function. A typical measure of response is the ACR20, which is a binary outcome assessing whether patients improve by at least 20% in a constellation of RA symptoms. A typical measure of function is the Health Assessment Questionnaire-Disability Index (HAQ-DI), which is a patient-reported outcome that measures disease-associated disability. Although the HAQ-DI is an ordinal outcome ranging from 0 to 3, it is commonly analyzed as a continuous endpoint in terms of change from baseline.

Most new biologics for MTX-IR patients are tested in combination with stable background MTX therapy. In some trials, the biologic may be tested in combination with MTX and as monotherapy. After randomization, several changes in treatment may be anticipated – some planned and some unplanned. Most trials in MTX-IR patients have a planned assessment of minimal required response to treatment, for example, ≥20% improvement from baseline in both tender and swollen joint counts at a specific time point. Patients who do not have this minimal improvement are offered rescue therapy for ethical reasons. Rescue may involve adjustments to background therapy (e.g., an increase of MTX dose), addition of other cDMARDs such as sulfasalazine or hydroxychloroquine, increase in NSAID or prednisone dose, change in NSAID, or new NSAID or prednisone start, intraarticular corticosteroid administration, or any combination of the preceding. Moreover, RA studies often included as rescue an "escape" (or "step-up") therapy with the investigational product, where patients randomized to placebo are switched to the experimental drug and patients randomized to a lower dose of experimental drug are switched to a higher dose. These escape treatment switches are typically implemented in a blinded manner and triggered by a protocol-defined requirement for rescue, such as the minimal required response in swollen and tender joint count mentioned above. Escape therapy may confound and complicate evaluations of some estimands at time points after its initiation but allows for longer duration of exposures to investigational product (e.g., when placebo-treated patients are switched to investigational product) to supplement the safety database.

If escape therapy to a higher dose of investigational product occurs, it may also help answer whether dose titration is an option in clinical practice. However, historically in RA trials, treatment effects evaluated per the primary efficacy objectives typically excluded the confounding effects of rescue. Switching to other nonstudy biologic agents typically is not part of the protocol-allowed rescue but may occur because of physician and patient decision. These post-randomization treatment changes constitute ICEs and require careful consideration in the estimand definition.

5.4.2 Trial Description

We now discuss a hypothetical trial with design elements that resemble a Phase 3 study of golimumab, GO-FORWARD (Keystone et al., 2009). Our example trial is a 24-week double-blind, placebo-controlled Phase 3 trial in MTX-IR patients evaluating an experimental biologic agent. The four treatment arms were (1) placebo injections plus MTX capsules, (2) experimental drug injections at high dose as monotherapy (i.e., with placebo capsules instead of MTX), (3) experimental drug injections at low dose plus MTX capsules, or (4) experimental drug injections at high dose plus MTX capsules. Injections were administered every 4 weeks.

The primary efficacy evaluation was based on co-primary endpoints: ACR20 at Week 14 and change from baseline to Week 24 in HAQ-DI score. Although HAQ-DI can be measured at earlier time points, function typically follows symptom improvement and may continue to increase over time and, therefore, later assessments may be more meaningful (Radner et al., 2015).

5.4.3 Estimand for RA Study Design 1

We first consider a study design and a primary efficacy objective mimicking the GO-FORWARD study as inferred from the description of methods in published material. We define an estimand in that context following the trial design process chart presented in Chapter 3.

1a. *Identify decision maker*: The primary decision makers are the regulatory agencies.

1b. *Define objective*: The primary objective is to assess the benefit of the experimental drug compared to placebo at specified time points in MTX-IR patients when taken as an add-on treatment without any modifications of therapy post-randomization. A decision to be made is whether to grant marketing authorization approval.

2a. *Identify possible ICEs*: Table 5.3 notes treatment sequences that may occur in this type of RA trial. Ideally, all patients would stay on the randomized treatment (Z) through the end of trial (Week 24). As in the GO-FORWARD study, the need for rescue is assessed at Week 16 based on the predefined criteria, and the rescue offered as part of the study is the escape therapy discussed above. All the treatment modifications implemented as part of this escape therapy are done in a double-blind manner. Escape therapy is denoted by "E" in Table 5.3. In addition to the planned rescue, it is anticipated that investigators may modify the treatment per clinical judgment. For example,

TABLE 5.3

Anticipated Treatment Regimens in a Trial of RA in MTX-IR Patients

	Week 4	Week 8	Week 12	Week 16	Week 20	Week 24	Treatment Regimen over 24 Weeks
1	Z	Z	Z	Z	Z	Z	Z
2	Z	Z	O	O	O	O	$Z \rightarrow O$
3	Z	Z	Z	E	E	E	$Z \rightarrow E$
4	Z	P(i)	P(i)	P(i)	P(i)	P(i)	$Z \rightarrow P(i)$
5	Z	C(i)	C(i)	C(i)	C(i)	C(i)	$Z \rightarrow C(i)$
6	Z	P(ii)	P(ii)	P(ii)	P(ii)	P(ii)	$Z \rightarrow P(ii)$
7	Z	Z	Z	E	P(ii)	P(ii)	$Z \rightarrow E \rightarrow P(ii)$
8	Z	Z	Z	E+C(i)	E+C(i)	E+C(i)	$Z \rightarrow E + C(i)$

Note: Z = randomized treatment as add-on to prestudy MTX regimen; O = background prestudy MTX regimen only; P = postdiscontinuation of randomized treatment; C = concomitantly with the randomized treatment; E = escape treatment; (i) = increased dose of MTX above the baseline dose for treatment of RA, new conventional disease-modifying antirheumatic drugs (DMARDs) or systemic immunosuppressive agents, treatment with oral corticosteroids for RA (new or dose above the baseline dose), or intravenous or intramuscular administration of corticosteroids for RA; (ii) = nonstudy biologic agents for RA.

investigators could (i) increase dose of MTX above the baseline dose or initiate new cDMARDs or systemic immunosuppressive agents or modify treatment with oral, intravenous, or intramuscular corticosteroids for RA (new drug or dose of existing drug above the baseline dose) and (ii) initiate nonstudy biologic agents for RA.

In our example, treatment modifications (i) may be initiated either after a permanent discontinuation from the study treatment (denoted by "P" in Table 5.3) or concomitantly with the randomized treatment (denoted by "C" in Table 5.3). Initiation of a new nonstudy biologic would normally occur after the permanent discontinuation of the study treatment in this patient population. Some patients may discontinue the randomized treatment and not initiate any new treatment before the time point of interest, that is, remain on their background therapy only (denoted by "O" in Table 5.3). Other patients may initiate the escape therapy, but then also initiate other changes (row 7 in Table 5.3).

We do not consider an exhaustive list of possibilities for timing of treatment modifications – the important point is that they may occur before the time points at which efficacy is evaluated, and therefore the changes in treatment affect the interpretation of treatment effects even if all patients are fully assessed through Week 24. Note that one of the two co-primary efficacy endpoints, HAQ-DI, is measured at Week 24 – that is, after escape therapy could be initiated.

All treatment sequences in Table 5.3 except scenario 1 are ICEs for which a strategy needs to be specified as part of the estimand definition.

2b. *Define treatment regimen under evaluation*: The treatment regimen under evaluation is the randomized treatment taken for up to 24 weeks without any adjustments to the background therapy, with allowance for early discontinuation due to reasons other than lack of efficacy. This is scenario 1 in Table 5.3.

2c. *Define estimand*: The primary estimand that mimics the one we infer from the GO-FORWARD study is:

    a. The treatment effect for the population of adult patients with active RA despite MTX therapy (MTX-IR) as defined by the protocol inclusion/exclusion criteria.

    b. Efficacy is measured using two co-primary endpoints: ACR20 at Week 14 and HAQ-DI score at Week 24.

    c. To handle ICEs: For patients who prematurely discontinue the randomized treatment for reasons other than lack of efficacy and do not initiate any adjustments to the background therapy, observed outcomes at the designated time points provide nonconfounded evidence for the effect of the treatment regimen under evaluation. Therefore, the treatment policy strategy will be used for this ICE.

For patients who initiate a protocol-defined escape therapy, a hypothetical strategy will be used to estimate what the treatment effect would have been at the designated time point if patients did not receive the escape therapy and continued their randomized treatment. It is assumed that such patients would remain in the same condition as prior to escape initiation through the time point of interest.

Patients who prematurely discontinue the randomized treatment for lack of efficacy, adverse events, or initiate any treatment adjustments other than the protocol-defined escape therapy will be considered treatment failures at the designated time points after the start of such treatment changes. Therefore, the composite strategy is used for these ICEs. This approach has traditionally been referred to as non-responder imputation (NRI). However, NRI is a misnomer in that values are not imputed because no values are missing; rather, patients are defined as treatment success or failure. Any patient who discontinues for lack of efficacy or requires nonescape treatment adjustment is classified as a treatment failure regardless of ACR20 outcomes, other patients' outcomes are based on ACR20 response.

d. The summary measures, the proportion of treatment successes at Week 14 and median changes from baseline to Week 24 in HAQ-DI, will be compared between the treatment groups. The treatment policy strategy for ICEs described in C1 accounts for imperfect compliance, including early discontinuation of study treatment. A determination of the primary reason for discontinuation, however, could be based on a subjective judgment of patients and/or investigators and not on formal criteria such as ACR20. It is, therefore, important to provide clear guidance in the protocol for determining the primary reason for discontinuation and to monitor these data during the study.

The strategy for handling ICEs fulfill the objective of estimating the effect of randomized treatment without any confounding by the effect of other medications. In general, assumptions about the data unavailable for some patients with ICEs under the hypothetical scenario may not need to be stated in the estimand definition and deferred to the description of the estimator, if there are some patients in the trial who are expected to follow the hypothetical scenario. It is implied that such patients will be used as a reference group in a statistical model to predict unavailable outcomes. In the current context, however, all patients requiring rescue as per the protocol-defined criteria are expected to initiate the escape rescue, and therefore no reference group can be identified. In this case, it is beneficial to state some clinical assumptions as the basis to estimate the outcomes under the hypothetical strategy so that an estimator can be aligned with these clinical assumptions.

A composite strategy used for all other ICEs besides escape (rescue) assume that these events represent treatment failure and that continuing with the randomized treatment alone would provide no chance of subsequent improvement.

3a. *Data useful for estimand*: Usable data that should be collected for this estimand are measurements used in the ACR20 and HAQ-DI at baseline and through Week 24 for all patients except those with ICEs for which a composite strategy is used.

In the GO-FORWARD trial, patients who prematurely discontinued the randomized treatment continued to be evaluated for safety and selected efficacy assessments for 4 months after the last dose of study treatment. In more recent RA trials, a typical regulatory recommendation is to continue study participation (with efficacy and safety evaluations) for the duration of the double-blind period, with possibly limiting the assessments to the essential ones, and these additional data are used for supportive analyses.

3b. *Patient retention strategy*: Study personnel should work with patients to encourage their continued participation in the study, including in cases of premature discontinuation of the randomized treatment. In discontinuations for reasons other than lack of efficacy, these follow-up data are critical because they are used for the primary estimand; in other cases, follow-up data are useful for supportive estimands.

4a. *Main estimator*: For analysis of treatment success/failure at Week 14, standard methods for estimation of proportions and their differences should be used. Examples include logistic regression, chi-square, Fisher's Exact, or Cochran–Mantel–Haenszel tests. For analysis of changes from baseline to Week 24 in HAQ-DI scores, a rank-based method, for example, Wilcoxon rank sum test, can be used. Patients with ICEs who are considered treatment failures for this estimand are assigned the worst rank (see Chapter 10 for a discussion of estimators consistent with composite strategies, including rank-based estimators and trimmed mean approaches).

The estimator also must accommodate patients who utilize escape rescue treatment and need to be handled with a hypothetical strategy – assuming such patients would remain in the same condition as prior to escape. In the GO-FORWARD study, this was implemented using a last observation carried forward (LOCF) single imputation approach. Single imputation can lead to underestimation of variance, but an LOCF-like approach can also be implemented using multiple imputation (see Chapter 19 for details on multiple imputation).

4b. *Missing data assumption*: Data may be missing intermittently if a patient without any ICEs miss assessments or when patients withdraw from the study due to reasons other than efficacy and adverse events. In these cases, it is reasonable to assume that the missing outcomes would be similar to those of patients with similar baseline and previous post-baseline values in their treatment group (the MAR assumption, see Chapter 15). Multiple imputation can be used to impute these missing values. The amount of missing data should be minimal in a well-executed study. Unobserved outcomes of patients with ICEs are assumed to be similar to their outcomes prior to escape initiation.

4c. *Sensitivity estimators*. To assess sensitivity to missing data assumptions, a more conservative assumption is often used where all patients with missing/unobserved data as described above are considered as treatment failures. Delta adjustment/tipping point analyses can also be performed (see Chapter 14 for a discussion on sensitivity analyses, and Chapter 23 for details on delta adjustment).

4d. *Sample size*: Sample size requirements should be based on assumptions that incorporate the likely rates of ICEs described in C of the estimand definition above and an impact on the overall treatment effect due to patients with these events.

5.4.4 Estimand 2 for RA Study Design 2

To illustrate a broader range of possibilities, we define an estimand for a different context and describe a different study design to accommodate this new context. The key differences are protocol-defined adjustments to the background therapy that are part of the rescue therapy; patients requiring rescue are randomized to either initiate the escape therapy or remain on the current regimen with the modified background therapy; all premature discontinuations of the randomized treatment are considered treatment failures regardless of discontinuation reason to avoid relying on subjective judgments. Following the study development process chart for this scenario,

1a. *Identify decision maker*: The primary decision makers are the regulatory agencies.

1b. *Define objective*: The primary objective is to assess benefit of the experimental drug compared to placebo at specified time points in MTX-IR patients when taken as an add-on treatment allowing for specific adjustments to the background therapy that are commonly undertaken in clinical practice. The decision to be made is whether to grant marketing authorization approval.

2a. *Identify possible ICEs*: The details of anticipated ICEs listed in Table 5.3 need to be refined to split the treatment changes in "(i)" into adjustments to the background therapy that are part of the protocol-allowed rescue therapy and those that are not. Need for not-allowed adjustments is considered treatment failure. The allowed adjustments are prespecified in the protocol and could include an increase of MTX dose or change in route of administration, addition of other cDMARDs such as sulfasalazine or hydroxychloroquine, new NSAID or change in NSAID dose, modifications of corticosteroids use, or any combination of these. Prespecification of allowable adjustments enables inferences about the specific treatment regimen. For the primary evaluation of efficacy, the confounding effect of treatment switching on HAQ-DI at Week 24 in patients who initiate escape therapy must be accounted for. The hypothetical strategy for what would happen if patients continued with the treatment regimen under evaluation without the escape therapy can again be employed. However, to implement this strategy in a more robust manner that does not rely solely on assumptions, patients requiring rescue are randomized in a blinded manner to either initiate escape or not so that data can be collected from some rescued patients that followed the hypothesized scenario while testing the clinically useful hypothesis regarding whether adding rescue treatment is beneficial. In line with recent regulatory recommendations, all patients meeting requirement of rescue initiate protocol-allowed changes in their background therapy regardless of whether they are randomized to escape or not.

2b. *Define treatment regimen under evaluation*: The treatment regimen under evaluation is the randomized treatment taken for up to 24 weeks, with possible protocol-defined adjustments to the background therapy as rescue.

2c. *Define estimand*: Elements of the estimand definition are like those specified for RA study design 1, except for element C, that is, handling of ICEs.

C1. For patients who require rescue and have background therapy adjustment as allowed per protocol, without initiating an escape therapy, observed outcomes at the designated time points provide nonconfounded evidence for the effect of the treatment regimen under evaluation. Therefore, the treatment policy strategy is used for these ICEs.

C2. For patients who initiate a protocol-defined escape therapy, a hypothetical strategy is used to estimate what the treatment effect would have been at the designated time points if patients did not receive the escape therapy and continued their randomized treatment, with protocol-allowed adjustments to the background therapy.

C3. Patients who prematurely discontinue the randomized treatment for any reason or initiate any not-allowed treatment or adjustment in background therapy are considered treatment failures at and after the time points when the change in treatment regimen occurred. That is, a composite strategy is used for these ICEs.

Considerations for items *3a (data useful for estimand)* and *3b (patient retention strategy)* are similar as in the case of the RA study design 1.

4a. *Main estimator*: Analysis considerations for this estimand are like those discussed for the RA study design 1, except for handling of patients with ICEs described in C2. Rather than an LOCF approach, hypothetical outcomes under no escape therapy are modeled. Data from rescued patients who are randomized not to initiate the escape therapy are fit to a statistical model that is used with multiple imputation to estimate hypothetical outcomes under no-escape therapy for patients who were randomized to escape. This multiple imputation model should include baseline covariates and post-baseline assessments prior to rescue and can be implemented using reference-based imputation (see Chapter 22 for details on reference-based imputation).

4b. *Missing data assumption*: Considerations for this estimand are similar to RA study design 1, except for the assumption used with the hypothetical strategy. Patients with ICEs described in C2 are assumed to have similar efficacy outcomes as patients in their treatment group who also met conditions for rescue therapy and had their background therapy adjusted in the protocol-defined manner without receiving the escape therapy.

4c. *Sensitivity estimators*. Similar sensitivity analyses as mentioned for the RA study design 1 can be used.

4d. Sample size. Same as for design 1.

5.5 Summary

This chapter illustrated examples of defining estimands in real-life settings consistent with the concepts outlined in ICH E9(R1) and discussed in previous chapters. The three examples illustrated a variety of ICEs that can be anticipated in each setting, along with strategies that can be used to handle the ICEs consistent with the study objectives and the clinical context. The estimands chosen for these examples are not the only acceptable choices for their respective scenarios. The intent is to illustrate the process and key concepts rather than focus on justification of specific choices.

Emphasis was placed on following the study design process chart described in Chapter 3. Following the steps outlined in that process helps ensure alignment between objectives, estimands, data, and analysis, thereby fostering useful inferences. The ICH E9(R1) emphasized the importance of defining estimands before choosing estimators and defining estimands that are not overburdened by statistical details so that trialists from all backgrounds can understand and contribute to the estimand definition. Our examples demonstrated that this is feasible. However, in complex situations like the RA example, there will inevitably be greater detail than in simpler scenarios, such as example 1 with the PoC trial in MDD.

6

Causal Inference and Estimands

6.1 Introduction

The fundamental concepts for handling missing data in clinical trials, such as the taxonomy of missing data mechanisms (see Chapter 15), along with multiple imputation and likelihood-based analyses, trace to key works of Rubin (1976, 1978a). A critical difference between the public survey context on which Rubin's missing data concepts were initially focused and the clinical trial context is that surveys do not usually involve interventions (and therefore estimands are of a "representative" nature, for example, the gross national income of a country), whereas clinical trials focus on the effects of interventions. This difference may help explain why clinical trialists have adopted the methodology for handling missing data first and only recently focused on the need to clearly specify what effects of the intervention – the estimand – are the target of inference (Lipkovich, Ratitch, and Mallinckrodt, 2019).

As seen in Chapter 5, describing estimands using the four elements suggested by the ICH E9(R1) involves detailed verbal definitions but may still leave room for ambiguity. As with specifying equations or statistical models, the precise language of mathematics can help alleviate ambiguity among those familiar with the notation and terminology.

Causal inference methods, such as those in Greenland, Pearl, and Robins (1999), Pearl (2001), Rubin (1978b), and Robins (1986), were developed initially for observational data where confounding due to outcome-based treatment assignment and treatment change is common. Causal inference methods have also been applied to clinical trials, with initial application motivated by estimating treatment effects under noncompliance (Robins, 1998; Rubin, 1998). Utilization of causal inference methods grew with an increasing appreciation for the similarity between clinical trials and observational data due to various post-randomization events that effectively break the randomization in clinical trials (Lipkovich, Ratitch, and Mallinckrodt, 2019).

Although the problems caused by intercurrent events (ICEs) are common to both clinical trials and observational studies, a gap has existed between the clinical trials and causal inference literatures (Lipkovich, Ratitch, and Mallinckrodt, 2019). Prior to the NRC expert panel report on the prevention and treatment of missing data (NRC, 2010), the clinical trial literature on missing data rarely distinguished missing data caused by treatment discontinuation from switching to or adding rescue medications because common practice was to withdraw patients that required rescue from the study (Lipkovich, Ratitch, and Mallinckrodt, 2019).

However, the NRC report (NRC, 2010) advocated retaining patients and continuing to assess them regardless of adherence to the initially randomized treatments. Therefore, clinical trialists had to consider what to estimate and how to estimate it in the presence of discontinuation of or changes in treatment, a problem common in observational settings (Lipkovich, Ratitch, and Mallinckrodt, 2019).

Despite the similarity in challenges encountered in causal and clinical trial applications, clinical trialists have not fully leveraged causal language and concepts. One reason may be that elements of the causal framework, such as potential outcomes (POs), are unfamiliar to those with classical training in clinical trial statistics.

Nevertheless, causal language, such as POs, can render defining estimands easier. For example, when POs are defined, and the treatment contrast is specified at the patient level, the estimand is simply the expectation over the entire population. This approach is especially useful in helping to clarify the treatment regimen of interest and which ICEs are deviations from that regimen (Lipkovich, Ratitch, and Mallinckrodt, 2019).

6.2 Causal Framework for Estimands in Clinical Trials

6.2.1 Defining Potential Outcomes

Defining causal estimands requires defining causal effects. Clinical trial statisticians often receive little training in causal effects other than to appreciate that association does not imply causation. In conventional statistical models, causal estimands are often structural parameters (e.g., coefficients associated with treatment variables) conditional on other covariates. However, the parameter is causal only in settings of randomized or simulation experiments when all sources of confounding can be controlled. A researcher can pose a model for observational data and postulate that the coefficient for some variable is causal, but this does not yet define what is causal (Lipkovich, Ratitch, and Mallinckrodt, 2019).

Defining causal effects requires additional concepts outside traditional statistics, such as POs (Neyman, 1923; Rubin, 1978b) that will then lead to defining causal estimands. The PO framework has been criticized by some on philosophical and practical grounds (Dawid, 2000). Nevertheless, POs have pragmatic utility because they provide a unifying language to describe both experimental and observational settings. Importantly, POs facilitate defining causal effects *independently* from the treatment assignment mechanism (Lipkovich, Ratitch, and Mallinckrodt, 2019).

Let $Y_i(z), z = 0,1$ be the PO that would be realized if a randomly selected patient i, received treatment $Z = z$. The difference between $Y_i(z)$ and traditional random variables is that its realized values are conditional on variable Z that has values set through intervention or manipulation rather than being simply observed.

POs are defined with respect to all treatments under investigation, regardless of which treatment the patient was randomized to or took. For example, consider a patient who was randomized in a parallel group design to treatment $Z = 0$. The PO $Y_i(0)$ for this patient can be observed, whereas the PO $Y_i(1)$ cannot be observed, yet it is well defined in principle. In the clinical trial literature, such an outcome is often termed counterfactual because it represents the outcome of treatment $Z = 1$, which is counter to the fact that the treatment was $Z = 0$ (Lipkovich, Ratitch, and Mallinckrodt, 2019).

6.2.2 Counterfactual Outcomes and Potential Outcomes

The terms "counterfactual outcome" and "potential outcome" often are used interchangeably (Robins, 1998). Some prefer the term "potential outcomes" to emphasize that both $Y_i(1)$ and $Y_i(0)$ are well defined before the treatment assignment and both have a potential of being observed in a parallel group design. It can be useful to distinguish between counterfactual POs that can and cannot be observed. In so doing, it is important to keep separate the ideas of whether the probability of a PO is zero from whether interest exists in a PO. Unobserved POs and estimands associated with them can be relevant and provide context to the observed POs and estimands. No single clinical trial can address all the relevant scientific and medical questions regarding an intervention. Therefore, some relevant POs will not be observed in a trial and some relevant estimands will not be estimable from the data alone.

For example, if a patient discontinues treatment and drops out of the study interest may be in the PO that would have been observed had the patient continued the initially randomized study drug or continued in the study on rescue medication. In many scenarios, these POs have probability > 0 of occurring. Alternatively, interest may be in the POs of the same patient had they continued in the trial untreated. In many scenarios, it is unethical to leave patients untreated and the probability of observing this PO in the trial is zero.

Another way to consider this is via the methods for dealing with ICEs discussed in Chapter 4. Hypothetical approaches to dealing with ICEs can involve POs with probability equal to zero of being observed in the actual trial.

6.3 Using Potential Outcomes to Define Estimands

6.3.1 Defining Estimands

The following text, adapted from Lipkovich, Ratitch, and Mallinckrodt (2019), describes how to define estimands using POs. POs relate to observed outcomes Y

$$Y_i = Y_i(1)Z_i + Y_i(0)(1-Z_i).$$

Thus, POs under a certain treatment Z are consistent with observed outcomes for those patients who took that treatment. This is referred to in the causal literature as consistency. Importantly, the relation is not an identity, but an assumption that is implied by a more general "stable unit treatment value assumption" (SUTVA) (Little and Rubin, 2000).

Under SUTVA, the treatment status of any patient does not affect the POs of other patients. An example of a violation of this assumption is POs associated with a vaccine for a contagious disease that may depend on whether other people in the community had received a similar vaccine.

Another implication of SUTVA is that no hidden variation in treatments exist sometimes referred to as the "no-multiple-versions-of-treatment assumption" (VanderWeele and Hernán, 2013). Unaccounted-for variation can arise from unmodeled effects, such as if some patients took rescue medication that influences POs under their main treatment, but the analysis does not account for the use of rescue. This underscores the need for explicitly defining the treatment or treatment regimens under investigation as noted in Ratitch et al. (2019a) so that the deviations from this regimen are clear – and hence not hidden.

Causal estimands can be defined in terms of expectations of POs. In the context of a clinical trial, the PO $Y(z)$ and the observed Y can, for example, be defined for a specific time or time interval (e.g., at week 24). The average treatment effect can be defined in terms of expectation of the *individual* treatment effect, that is, expected difference between POs (say for a continuous or binary Y) as

$$\delta = E\big(Y_i(1) - Y_i(0)\big).$$

Throughout the rest of this chapter, the patient index is dropped to simplify notation. Note that δ has a causal interpretation regardless of whether treatment is assigned randomly or not.

The causal estimand δ can be estimated from a randomized trial because

$$\delta = E\big(Y(1) - Y(0)\big) = E\big(Y(1) - E\big(Y(0)\big)\big)$$

$$= \big(EY(1)|Z = 1\big) - E\big(Y(0)|Z = 0\big)$$

$$= E(Y|Z = 1) - E(Y|Z = 0)$$

The second line uses the independence of POs from randomized treatment assignment, that is, independence of Z from $Y(0)$ and $Y(1)$. This is not equivalent to Z being independent of Y, which is in general not true. Here, it means that the distribution of a PO had patients been assigned to active treatment does not depend on whether a patient was in actual fact assigned to active treatment or placebo. It does not mean that the observed outcome $Y \equiv Y(Z)$ associated with actual treatment assignment is independent of Z.

The last step substitutes POs with fully observable outcomes Y (under SUTVA) under the assigned treatment because patients in the randomized treatment groups are exchangeable and the analysis proceeds using standard statistical methods for estimating expected values of a random variable.

In the absence of randomization, it may still be possible to estimate δ, if it can be assumed that no unmeasured confounding exists. That is, the POs are independent of treatment assignment Z after conditioning on measured confounders X (e.g., baseline covariates that are used by physicians in making treatment decisions in clinical practice). In this case, the average treatment effect can again be estimated because

$$E\big(Y(1)\big) = E\big(E\big(Y(1)|X\big)\big) = E\big(E\big(Y(1)|X, Z = 1\big)\big) = E\big(E(Y|X, Z = 1)\big)$$

(where the outer expectation is taken with respect to X) and similarly for $E(Y(0))$.

The assumption of no unmeasured confounding in causal inference has analogy to the missing at random (MAR) assumption for missing data in clinical trial contexts (see Chapter 15). Consider the ICE of early patient discontinuation. Under MAR, confounding from the missing data that arose from early discontinuation can be removed by conditioning on all relevant observed factors that fully account for the early discontinuation. It is useful to consider early discontinuation as a change in treatment from randomized treatment to no treatment. In this context, MAR implies adjustment for confounders of early discontinuation, and early discontinuation defines a treatment regimen. This is conceptually akin to adjustment for confounders of treatment in nonrandomized experiments.

The causal inference framework distinguishes interventions from outcomes. Interventions are acts of decision, even if medically necessary. Outcomes are events. For example, if a patient experiences a severe adverse event and discontinues study medication, the adverse event is the outcome (S) and discontinuation is the intervention (R). The distinction between outcomes and interventions is central to the causal framework developed by Robins and colleagues (Robins, 1998; Robins et al., 2000; Robins and Hernán, 2009) for evaluating estimands in longitudinal data with time-dependent confounding (see also the tutorial by Daniel et al., 2013). In the most general case, initial (or randomized) treatment Z affects various outcomes S, which in turn affect subsequent changes in treatment R that (along with all the previous) affects subsequent outcomes Y.

Here, S is a time-dependent confounder as illustrated in Figure 6.1. The notion of time-dependent confounding through intermediate outcomes S is important because S is both a confounder (similar to baseline covariates in an observational trial that affect initial treatment assignment) and a mediator of the effect of the initial treatment Z on the outcome Y. This simple scheme that contains a single pair of post-baseline outcomes S and treatment R can be generalized to multiple outcomes and treatment changes at different times.

Some of the causal estimands that can be defined are as follows:

- The effect of initial treatment Z (total and partitioned into direct effects and indirect effects mediated via S or R)
- The joint effect of Z and R (free of confounding by S)
- The effect of Z conditioned on POs of intermediate outcomes {S(0), S(1)} or potential treatments {R(0), R(1)}
- The effect of a dynamic treatment regimen in terms of Z and R, as a function of S

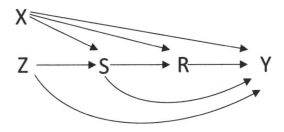

FIGURE 6.1
Causal diagram showing relationships between baseline covariates X, randomized treatment Z, intercurrent events S (outcome) and R (intervention), and the ultimate outcome Y.

6.3.2 Specifying Treatment Changes

Lipkovich, Ratitch, and Mallinckrodt (2019) distinguished post-randomization outcomes and changes in treatment that may confound inference about causal effects. Interest may be in intermediate outcomes (S) when the goal is to explain a mechanism of action, which may be mediated through intermediate outcomes. However, in confirmatory clinical trials, interest focuses on post-randomization treatments (R) for the main objective of attributing the causal effects of the interventions. Alternatively, interest may be in the treatment effect defined by principal strata based on outcome S and not on treatment R.

In some clinical trials, dosing adjustments are allowed, and in other trials, dosing is fixed. Regardless, the treatments taken may change in various ways, such as adding or switching to a new treatment or discontinuation of previous treatments (either background or study treatment). These changes may be mandated or discretionary. Treatment changes are influenced by the evolving outcomes that result from the treatment received so far. Therefore, it is of interest to consider estimands and POs wherein assigned treatment may (dynamically) change during the trial in response to patient outcomes.

The causal inference literature distinguishes causal effects that can be inferred from time-varying interventions (Lipkovich, Ratitch, and Mallinckrodt, 2019). Using Figure 6.1, it is useful to assume the initial treatment assignment is to an intervention z from a set of study treatments Z. Prior to assessing the outcome of interest Y, patients could receive another intervention $r \in R$, where the set R may include the same treatments as Z (i.e., to continue the initially assigned treatment or change it).

Consider the *total* and *joint* effect of Z on outcome Y as defined by Daniel et al. (2013). The total effect of the initial treatment Z on Y includes the effect of R to the degree that R has been influenced by Z, and the intermediate outcomes that followed Z. That is, the total effect includes the direct effect of Z and the effect that is mediated by the later treatment R. The joint effect of the treatment history is defined as a collection of direct effects, that is, the direct effect of Z unmediated by R and a direct effect of R. These distinctions are important and have implications in clinical practice. As an example, consider treatment Z whose initial effect is large but short lived. The total effect of Z will be dominated by effects mediated via R (Lipkovich, Ratitch, and Mallinckrodt, 2019).

The total effect is what would be targeted by a treatment policy strategy, including the so-called pure intent-to-treat (ITT) estimand. Joint effects describe and compare POs of specific (or fixed) time-varying treatment sequences and in doing so provide clarity on the causal effects of each component.

Dynamic treatment regimens (also often referred to as "dynamic treatment regimes") differ from the joint effects of prespecified treatment sequences. In contrast to the pure ITT/treatment policy strategy where deviations from assigned treatment are not relevant, dynamic treatment regimens entail precisely specified rules for treatment assignment in response to evolving patient outcomes. A PO for a dynamic treatment regimen is an outcome (response) for an arbitrary patient had that patient (possibly contrary to fact) been given treatment according to a *specific* treatment regimen (Moodie et al., 2007).

Lipkovich, Ratitch, and Mallinckrodt (2019) provided the following examples of dynamic treatment regimens:

1. Patients continue treatment for t hours/days/weeks or until occurrence of an event that necessitates treatment termination, whichever occurs first. This regimen is sometimes called treatment duration policy. See Johnson and Tsiatis (2005) for an example that studies the effect of variable treatment duration in an infusion trial.

2. Patients continue treatment for t hours/days/weeks until "a rescue condition" occurs that leads to initiation of a rescue treatment or escalation of dose.

3. Similar to (2), where rescue criteria are met, but rescue is not provided. This regimen may be purely hypothetical if rescue must be provided due to the seriousness of the condition and where rescue is always available, or this regimen may be realistic if in some conditions, based on considerations outside of the primary rescue criteria, rescue may not be provided or may be unavailable.

4. A regimen where the dose can be adjusted up or down at any time in response to lack of efficacy or safety/tolerability.

Example 3 is a treatment regimen that may not be included in the trial for ethical reasons. However, the POs may still be meaningful for public policy makers to evaluate the effect that can be attributed to the initial treatment free from the confounding effects of rescue. This is an example of the point made in the previous section that being counterfactual within the context of a specific clinical trial does not mean the outcome is not relevant.

Assessing the various treatment effects requires formal definition of the POs associated with the initial randomized treatment. Moreover, ICEs must be defined as either post-randomization outcomes S or treatment changes R, although interest will typically focus on the latter. Lipkovich, Ratitch, and Mallinckrodt (2019) defined POs as a function of two arguments, randomized treatment Z and ICEs S, where S can refer to either outcome or treatment change, which we refer to with R when relevant, $Y(z,s)$.

Assume $S = 0,1$, although in principle S can be ordinal or continuous. That is, $Y(z,s)$ is a PO that would be observed if an arbitrary patient were initially assigned treatment $Z = z$ and experienced an ICE $S = s$ (even if contrary to fact). If the ICE is change of treatment R, then $Y(z,r)$ is the PO for a random patient who would be treated with $Z = z$ and $R = r$, even if contrary to fact. The joint effect of Z and R can be unambiguously expressed as a contrast in expectations of this PO (Lipkovich, Ratitch, and Mallinckrodt, 2019).

As a shorthand, $Y(z, S(z)) \equiv Y(z)$, where $S(z)$ is the level of the ICE if a patient was assigned to treatment z. This is sometimes termed the composition assumption (VanderWeele and Vansteelandt, 2009), which essentially entails that intervention on S is noninvasive. In other words, the PO intervening to set $Z = z$ is equal to the PO intervening to set Z to z and to set S to the value it would have been if $Z = z$. Similarly, $Y(z, R(z)) \equiv Y(z)$. The total effect of Z is then expressed as a contrast in expectation of these POs (Lipkovich, Ratitch, and Mallinckrodt, 2019).

The PO associated with a dynamic regimen (rule) R is an outcome $Y(R)$ that would have been observed if the dynamic regimen was applied to a randomly selected patient. Note that here R is a function returning treatment based on post-randomization outcomes for that patient (e.g., S and/or other measures). This simplified notation is for a single outcome but can be extended to repeated measures (Lipkovich, Ratitch, and Mallinckrodt, 2019).

Example (1) above involves a *dynamic treatment duration regimen* of up to t time units. Define potential time to treatment terminating events for patients randomized to control and active treatment, as $t_{dr}(0)$ and $t_{dr}(1)$, respectively, assuming the trial continues indefinitely. The potential time to treatment terminating events or end of study period t, is $\tilde{t}(z) = \min(t, t_{dr}(z)), z = 0, 1$. Let PO at arbitrary time u be $Y_u(z), z = 0, 1$. Then POs associated with dynamic regimens are $Y_{\tilde{t}(z)}(z), z = 0, 1$. Interest may be in estimating the expected PO at the end of treatment for each treatment as a function of treatment/study duration t, $E[Y_{\tilde{t}(z)}(z)] = m(t, z)$ (Lipkovich, Ratitch, and Mallinckrodt, 2019).

For examples (2) and (3) above, consider a trial where $R = 1$ if patients receive rescue medication and $R = 0$ if not receiving rescue medication. Assume the protocol specifies that rescue treatment is initiated if patients meet specific, predefined rescue criteria after initiation of the initial treatment. The rescue criteria are based on the intermediate outcome, L: for example, $R \equiv g_\theta(L) = I(L > \theta)$, where $g_\theta(\cdot)$ is a treatment regimen function, $I(\cdot)$ is an indicator function, and θ is a clinically relevant cutoff in symptom severity that defines the need for rescue (e.g., relapse) and lower values of L represent better outcomes (Lipkovich, Ratitch, and Mallinckrodt, 2019).

Therefore, in example (2), interest is in POs $Y(1, g_\theta(L(1)))$ and $Y(0, g_\theta(L(0)))$. The first argument of the PO is the randomized treatment, and the second is the ICE that results in the initiation of rescue therapy. In example (3), interest may be in $Y(1,0)$ and $Y(0,0)$, where the second argument set to 0 indicates that patients would NOT receive rescue regardless of whether the patients meet the rescue criterion (Lipkovich, Ratitch, and Mallinckrodt, 2019).

Some estimands can be described using dynamic regimen language to make their formulation more explicit. Dynamic regimen language is particularly useful when the regimen can be precisely stated and therefore described as a function $g_\theta(L)$ that can take a complex multiparameter form, with both θ and L being vector valued (Lipkovich, Ratitch, and Mallinckrodt, 2019). In observational data, treatment switches are typically based on subjective criteria, such as physician and patient opinion. In these instances, a family of estimands with a varying parameter, such as the threshold θ, can be estimated and an optimal regimen identified from the data.

In randomized clinical trials, the protocols are traditionally more restrictive, leaving less freedom for investigators/physicians when making decisions. If estimating the optimal regimen is the primary objective, trials can be designed as sequential multiple assignment randomized trials (SMART) to facilitate estimation of the effects of dynamic regimens and thereby identify the optimal regimens (Murphy, 2003, 2005; Moodie et al., 2007).

As with estimands, it is important that POs are meaningful and interpretable. Being able to define a PO does not mean it is relevant. Moreover, just as the relevance of estimands varies according to the clinical context, so does the relevance of POs. It is sometimes relevant to consider the POs patients would have had if they had not received rescue medication, counter to the fact that they did receive rescue, or it may be relevant to consider the POs patients would have had if they had not discontinued study medication.

It is useful to consider the ICE of adherence. The causal literature often refers to compliance in the same context as clinical trialists refer to adherence. Code adherence to active treatment and placebo as actual treatment $R \in \{0,1\}$. In causal inference notation, the PO for patient i who was randomized to active treatment and was then noncompliant is $Y_i(1,0)$. The first argument indicates the randomization group and the second argument the actual treatment at the time point of interest. A key consideration is whether the PO $Y_i(1,1)$ is meaningful given $Y_i(1,0)$ was observed. Here, it is useful to adopt the suggestion of VanderWeele (2011) to view the plausibility of hypothetical interventions and counterfactuals as a spectrum – some are more reasonable to entertain than others (Lipkovich, Ratitch, and Mallinckrodt, 2019).

To illustrate the utility of the degree of counterfactualness, consider patients in the above example who were noncompliant: (1) due to reasons unrelated to treatment, such as relocations making it impossible to continue participation in the trial; (2) due to adverse events that caused some but not all patients who experienced events of similar severity to discontinue; or (3) per protocol, mandated to discontinue. Scenario 3 is the most counterfactual because the probability of any patient continuing given these outcomes is zero. Scenario 1 is the least counterfactual because it is easy and relevant to envision other scenarios where these patients would have been adherent. Scenario 2 falls in between. It may be relevant to consider POs under adherence given that nonadherence was not certain. Moreover, when making decisions about whether to continue treatment or not, it is necessary to understand what the outcome would (likely) be if the patient continued. And as noted elsewhere, the counterfactual outcomes could be relevant if, for example, adherence in actual clinical practice differs meaningfully from that in the clinical trial where patients are blinded to medication and might be on placebo (Mallinckrodt, Molenberghs, and Rathmann, 2017).

6.4 Examples of Defining Estimands in the Potential Outcome Framework

6.4.1 Introduction

In this section, the PO framework is used to define example estimands for each of the five strategies for dealing with ICEs outlined in the ICH E9(R1) Addendum, along with estimands for scenarios in which patients have changes in treatment. These examples were adapted from Lipkovich, Ratitch, and Mallinckrodt (2019).

As a general introduction, it may be relevant to consider the common example in the causal literature as compliance versus noncompliance. Compliance may mean different things depending on the context. For example, noncompliance may refer to discontinuation of randomized treatment, or intermittently missing some doses, changes in background medications, or adding or switching to rescue medications that are not prescribed in the protocol.

Let R be the treatment regimen under investigation. For compliant patients, $R(0) = 0, R(1) = 1$. As before, consider a clinical trial with two arms, active treatment and placebo. Extensions for multiple treatments or multiple doses are straight forward. Next, consider examples of the five strategies for dealing with ICEs defined in the ICH E9(R1) Addendum (ICH, 2019). The following examples illustrate how the PO framework can be applied to defining estimands. Neither these examples are an exhaustive list nor they are specific recommendations.

6.4.2 Treatment Policy Strategy

Treatment policy strategy estimands can be defined in the PO framework as $E[Y(1, R(1)) - Y(0, R(0))] \equiv E[Y(1) - Y(0)]$. The treatment policy strategy essentially assesses the effect of being randomized to an intervention. A treatment policy strategy can be applied to all ICEs or to a subset of them, with another strategy used for the remaining ICEs. Therefore, the treatment policy strategy evaluates the total effect of potentially several treatments that are part of the defined treatment regimen of interest.

6.4.3 Composite Strategy

The composite strategy for dealing with ICEs can be implemented in a variety of ways such that ICEs are part of the outcome of interest. It is useful to assume binary outcome Y, with $Y = 1$ indicating a poor outcome such as relapse or non-response. Patients with an ICE outcome $S = 1$, such as discontinuing study medication due to adverse events, are assigned a poor outcome $Y = 1$ regardless of status on other clinical outcomes. The PO can be redefined as $\tilde{Y}(z) = I(Y(z) \vee S(z)), z = 0,1$. The estimand is $E[\tilde{Y}(1) - \tilde{Y}(0)]$.

6.4.4 Hypothetical Strategy

One of the many hypothetical strategies is to estimate what outcomes would have been observed if ICEs had not occurred. Assume the ICE of interest is adherence to the initially randomized treatment. The estimand of interest is the average treatment effect in all randomized patients, assuming all patients adhered to treatment, even if some patients were not adherent (counterfactual). In this instance, the average treatment effect $ATE = E[Y(1,1) - Y(0,0))]$. This is one instance of a broader family of hypothetical estimands.

As another example, consider a hypothetical treatment regimen: Treat until time t or until a precluding ICE; for those with an ICE that precludes treatment with assigned medication, provide no treatment until t. Even though ethical considerations may prohibit no treatment, it may still be relevant to assess the effects at t due to the initial treatment, thereby hypothetically assessing what would have been observed if no treatment had been given after discontinuation. This PO can be expressed as (omitting for simplicity indexing by time) a mixture of POs $Y(z,z)(1 - S(z)) + Y(z,0)S(z), z \in \{0,1\}$, where $S = 1$ means the ICE. The estimand of interest is then a pattern-mixture estimand $E[Y(1,1)|S(1) = 0]Pr(S(1) = 0) + E[Y(1,0)|S(1) = 1]Pr(S(1) = 1) - E[Y(0,0)]$. This representation assumes that the PO for a patient who is not treated after discontinuation is drawn from the same population as those for placebo control, which may not always be realistic, such as when the control is an active treatment rather than placebo. Coding no treatment as "−1," the estimand can be expressed as follows:

$$E[Y(1,1)|S(1) = 0]Pr(S(1) = 0) + E[Y(1,-1)|S(1) = 1]Pr(S(1) = 1)$$

$$-E[Y(0,0)|S(0) = 0]Pr(S(0) = 0) - E[Y(0,-1)|S(0) = 1]Pr(S(0) = 1).$$

With these more complex situations, expressing the estimand in words can be cumbersome. The PO notation can be clearer for those with experience in mathematical expressions.

6.4.5 Principal Stratification Strategy

An alternative to assessing what would have happened if all patients adhered is to assess the average treatment effect in the subset of all randomized patients that would adhere to both (all) treatments. This effect is termed the local average treatment effect (LATE). This estimand can be specified in POs notation as $LATE = E[Y(1) - Y(0) | R(1) - R(0) = 1]$ (i.e., conditioning is on the event $R(1) - R(0) = 1$, which occurs only when $R(1) = 1, R(0) = 0$).

Alternatively, consider the principal stratification-based estimand of the treatment effect in patients that would adhere to the experimental intervention regardless of their ability to adhere to the control treatment. This estimand can be specified in PO notation as $E[Y(1) - Y(0) | R(1) = 1]$.

The LATE estimand was introduced by Angrist, Imbens, and Rubin (1996) and shown to be an instrumental variable estimand under certain assumptions. Imbens and Rubin (1997) called this estimand the "complier average causal effect" (CACE). This estimand is a special case of the principal stratification effect introduced in Frangakis and Rubin (2002).

From this broader perspective, principal stratification can be based on any criterion that addresses meaningful subpopulation.

As examples of the diversity in application, it is useful to consider Mehrotra, Li, and Gilbert (2006) who utilized an estimand that assessed the effect of vaccination on viral load in the principal stratum defined by patients that would become infected (post-randomization) regardless of whether they were randomized to the vaccine or placebo.

6.4.6 While-on-Treatment Strategy

While-on-treatment estimands assess treatment effects up to or at the last observation at or prior to time t, which marks the end of adherence to treatment, without regard for when that last observation was taken. This treatment effect can be defined in the PO framework as $E[Y_{i(1)}(1) - Y_{i(0)}(0)]$. Alternatively, different summaries of outcomes While-on-treatment can be considered, such as the average of all available values in a longitudinal sequence. The while-on-treatment strategy yields a direct treatment effect that does not allow time-varying components and focusses on the outcome up to the point where a change would occur.

6.4.7 Scenarios with Dynamic Treatment Regimens

The five strategies for dealing with ICEs in the ICH E9(R1) Addendum (ICH, 2019) lead to estimands that target either:

- Direct effects by defining a new and meaningful outcome or a new population
- Hypothetical effects that could be joint effects of initial and post-randomization treatment or pattern-mixture effects
- Estimands where a total effect is of interest and it need not be decomposed into direct effects

The ICH strategies do not cover the effects of dynamic treatment regimens – which may be confused with treatment policy (total effects) but are different. Therefore, consideration is now given to defining dynamic treatment regimens using POs.

Consider the example of estimands for the regimen where patients have some change in treatment, such as addition of rescue medication, that is based on prespecified rules. Let $R = 1$ for patients that meet prespecified criteria for rescue as quantified in the rule $g_\theta(L) \in 0,1$, where L incorporates intermediate outcomes. For example, $g_\theta(L) = I(L > \theta)$ as in Section 6.3.2; hence, the occurrence of the ICE that triggers rescue is given by $I(L > \theta)$. The following examples detail some of the estimands that could be considered.

The treatment effect for dynamic treatment regimen $g_\theta(L) \in \{0,1\}$ is a function of an intermediate continuous outcome L, $E[Y(1, g_\theta(L(1))) - Y(0, g_\theta(L(0)))]$. If all patients followed the rule associated with regimen $g_\theta(L)$, this estimand is the same as given by the treatment policy strategy $E[Y(1, R(1)) - Y(0, R(0))]$. If the rules for treatment changes, whether they result in dose adjustment, adding rescue medication, switching to rescue medication, and/or changes to background medication, are not deterministically specified or are not followed precisely, the treatment policy strategy tests a regimen that is a weighted average of a collection of treatment regimens (Lipkovich, Ratitch, and Mallinckrodt, 2019).

Consider the treatment effect for a (hypothetical) fixed treatment regimen where rescue medication was withheld (i.e., $R(1) = R(0) = 0$) regardless of whether patients meet rescue criteria: $E[Y(1,0) - Y(0,0)]$. This scenario implicitly assumes that rescue is added on to the initial treatment and interest is in what would happen if patients were prevented from receiving rescue. If interest were in an estimand where patients who meet rescue criteria would then not be treated, the dynamic regimen can be defined as follows: Let need for rescue be defined as before $I(L > \theta)$. Define a new treatment regimen $g_\theta(L)$ as a mapping of L and the initial treatment Z onto post-randomization treatment R, where $R = 1$ is experimental treatment and $R = 0$ is control, which in this scenario is no treatment. Therefore, $g_\theta(Z, L) = I(L \leq \theta)Z$. In words, continue assigned treatment for patients who did NOT meet rescue; continue and switch to no treatment for those who meet rescue criteria. The resulting estimand is $E\left[Y\left(1, g_\theta\left(1, L(1)\right)\right) - Y\left(0, g_\theta\left(0, L(0)\right)\right)\right]$.

When using hypothetical strategies, it is important to precisely define the scenario of interest, which typically means stating both what would and would not happen when certain conditions are met. In the previous example, it is necessary to stipulate whether patients who met rescue condition continue the previous randomized treatment or continue with no treatment.

6.4.8 Treatment of Missing Data

Typically, composite, while-on-treatment, and treatment policy strategies for dealing with ICEs result in no missing data for the primary outcome, although intermittent missing data may exist. However, in the principal stratification and hypothetical strategies, missing data should be anticipated. When defining estimands in the PO framework, consideration must be given to missing values. Even if all patients complete the trial, some POs will be missing because each patient has multiple POs, for example, POs on treatment and control, only one of which can be observed in a parallel group trial. The attractiveness of the PO framework is that when defining estimands there is no need to distinguish which POs are observed and which are missing. Typically, the estimand can be defined as an expectation over differences in POs if POs could be observed for each patient. Thus, the missing data problem is naturally and appropriately moved to the estimation stage (Lipkovich, Ratitch, and Mallinckrodt, 2019).

Typically, to estimate an expectation of the difference in POs, such as $E\left(Y(1) - Y(0)\right)$, the expectation over $Y(1), Y(0)$ (partially unobservable by design) must be replaced by the expectations over observable (by design) outcomes Y (as illustrated in Section 6.2.1). In the estimation stage, the need may exist to account for missing Ys. Therefore, the PO framework is consistent with the ICH Addendum by tackling the missing data problem after the estimand has been defined. That is, the scientific question drives the objectives and estimand, after which an analysis is chosen, as in contrast to choosing an analysis and then back solving for the estimand and objectives addressed by the analysis (Lipkovich, Ratitch, and Mallinckrodt, 2019).

It is important to stress that discontinuation of treatment or study discontinuation is not the same as missing outcome (see detailed discussion in Little and Kang, 2014). Discontinuing the study or the treatment can be considered a switch in treatment (to another available treatment or to no treatment) and as such can be a part of the estimand. Depending on the estimand, the outcome may or may not be treated as missing. In the treatment policy strategy, switches in treatment do not result in missing data, whereas in a hypothetical strategy where the estimand focuses on the effects of the initial treatment, data after switches in treatment are not relevant for estimating the parameters of interest and post-switch data are considered missing.

6.5 Summary

Clinical trialists have only recently focused on the need to clearly specify what effects of the intervention – the estimand – are the target of inference. Describing estimands can be tedious, involving lengthy verbal definitions that may still leave room for ambiguity. As with specifying equations or statistical models, the precise language of mathematics can help alleviate ambiguity among those familiar with the notation and terminology.

Causal inference methods that were developed initially for observational data may also have useful application in clinical trials where ICEs break the randomization, thereby inducing problems like those in nonrandomized studies. Moreover, causal language, such as POs, can make specifying treatment regimens and defining estimands easier. POs nomenclature can be used to specify estimands for each of the five strategies for dealing with ICEs.

7

Putting the Principles into Practice

7.1 Introduction

Previous chapters have detailed a variety of considerations and means to choose and specify estimands, along with details on handling intercurrent events (ICEs). These are new and therefore relatively unfamiliar topics that require considerable attention. A potential pitfall when dealing with new and nuanced considerations is becoming mired in the detail and losing sight of the big picture. The intent of this chapter is to pull back from the details covered in previous chapters to refocus on the most salient points – that is, to refocus on the big picture.

In clinical trials, patients may experience a variety of post-randomization events that are related to treatment and outcome that can break the causal link between the randomized treatments and the outcomes, thereby clouding inferences about the treatment effects. Historically, these post-randomization events were treated in the context of missing data. Fundamental principles in the analysis of incomplete data arose from sample survey applications where there were no interventions. Perhaps then, it is not altogether surprising that clinical trialists have only recently focused on the need to clearly specify what effects of the intervention – the estimand – are the target of inference.

Historically, estimands have not been consciously chosen and clearly specified. Rather, the target of inference usually had to be inferred from the analysis. Essentially, the choice of analysis dictated the scientific question to be addressed. Of course, the process should be reversed with the scientific question driving the analysis.

The ICH E9(R1) Addendum laid out a conceptual framework for choosing and specifying estimands where the scientific question drives the choice of estimand, study design, and data analysis. In so doing, the addendum expanded the discussion beyond missing data to more broadly discuss how to account for all ICEs. Choosing and specifying estimands and designing studies and analysis plans consistent with the estimand can be a tedious and difficult process. Therefore, a detailed and structured framework for putting the concepts into practice is useful.

The study development process chart presented in Figure 3.1 is a tool to help deal with these issues in a systematic manner. This process chart is a more detailed version that builds upon fundamental concepts of earlier process charts. The new process chart follows the same general path of earlier iterations in first choosing objectives, then defining estimands, and then matching the design and analyses (primary and sensitivity) to the objectives, estimands, and design. The new process chart provides additional details that are helpful in addressing how ICEs should be handled and is therefore consistent with the new ICH E9(R1) Addendum guidance.

The precise language of mathematics can help alleviate ambiguity in specifying estimands, at least among those familiar with the notation and terminology. For example, the potential outcome framework and nomenclature from causal inference can make specifying treatment regimens and defining estimands easier.

The authors credit James Bell as co-author through his contributions to the paper upon which this Chapter is based.

7.2 Overview of Process

The study development process chart begins with choosing objectives, that in turn inform choice of estimand. That is, the scientific question of interest drives choice of estimand and subsequently trial design and analysis. Clinical trials may have many stakeholders. The objectives and scientific questions of interest may vary for different stakeholders because they make different decisions. A comprehensive analysis plan may have to address the needs of regulators, health technology assessors, the general scientific community, prescribers, patients, and caregivers. Therefore, while it is necessary to specify a primary estimand, it is also necessary to specify secondary estimands that address the needs of secondary stakeholders.

With clear objectives, it is then possible to define the treatment regimen of interest and to define which post-randomization events (ICEs) are deviations from the regimen of interest. Although many specific estimands are conceptually possible, four general scenarios account for most of the options. These four scenarios are defined by two domains: whether inference is sought for the initially randomized treatment or regimens that involve the initially randomized treatments, and whether the inferences pertain to the effects (of the treatment or treatment regimen) if taken as directed (*de jure*/efficacy estimand) or as actually taken (*de facto*/effectiveness estimand). The appropriate study design, data to be collected, and the method of analysis can, and usually do, vary across these four general scenarios.

Given the diversity in clinical trial scenarios, no universally best estimand exists. Therefore, the choice of estimand is driven by the situation at hand – the scientific question. The optimal choice of estimand even for a single stakeholder can vary depending on:

- Phase: whether the trial is an exploratory, confirmatory, or post-marketing trial
- Disease state: acute versus chronic
- Type of intervention: symptomatic versus disease modifying
- Ethical considerations: need to provide rescue medication prior to observing the primary endpoint
- Setting: in-patient (or highly controlled administration of treatment) versus out-patient (or less controlled administration of treatment)
- Endpoint: efficacy endpoint versus safety endpoint.

Specifying estimands entails defining the population, the endpoint, the summary statistic, and how ICEs are to be handled. The first three components are familiar to clinical trialists because they have typically been specified in all trial protocols. The new aspect is the handling of ICEs. The ICH E9(R1) Addendum proposes five general strategies for dealing with ICEs (see Chapters 3 and 4).

Different types of ICEs may be handled with different strategies for different estimands, and multiple strategies can be employed for a single estimand. Each of these strategies has strengths and limitations in differing scenarios. These strategies also have implications for trial design and conduct, especially regarding what data need to be collected.

An important practical issue is how much detail on estimands needs to be specified in the protocol versus what can be included in a subsequent statistical analysis plan. Protocols could become excessively burdened with details on definitions and varied strategies for handling ICEs across a wide variety of estimands that might be required to address the needs of diverse stakeholders. Given the idiosyncratic nature of clinical research, a universal recommendation for what details to include in the protocol and which to leave for the statistical analysis plan is not practical. However, as a guiding principle, the protocol should clearly detail the primary estimand and how it is to be evaluated, including what data are needed because this influences data collection and therefore trial design. It is also necessary to provide at least high-level detail on secondary estimands because this can also influence data collection.

For example, we can consider a primary estimand that focuses on the effects of the initially randomized medications. This estimand does not require data after discontinuation of the initial medication or the initiation of rescue medication. However, if secondary estimands are based on the treatment policy strategy for dealing with ICEs, then it will be necessary to collect data after discontinuation of the initial medication and/or initiation of rescue. Therefore, high-level details on the treatment policy estimand are needed to justify and guide the collection of postdiscontinuation data. However, detailed specification on the handling of ICEs and analytic methods for all secondary estimands (i.e., exploratory endpoints) may not be needed in the protocol and could be included in the statistical analysis plan prior to unblinding – if these endpoints entail no additional design or data collection considerations beyond these already covered for other endpoints.

It is important to distinguish ICEs that are deviations from the intervention under assessment versus those that are not deviations from that intervention. For example, consider the use of rescue medication: if the intervention under assessment is a treatment regimen of the initial medication followed by rescue, then the change in treatment of adding or switching to rescue is part of the intervention and is not an analytic consideration because inferences apply to the regimen that includes rescue. If the intervention under assessment is the initial medication only, then the change in treatment to add or switch to rescue is not part of the intervention and use of rescue must be dealt with using one or more of the four strategies for dealing with ICEs along with an analysis appropriate for that situation.

In the treatment policy strategy, all ICEs handled by this strategy are irrelevant and therefore can be ignored in analyses. The other four strategies either define ICEs as an outcome or define what data are relevant in estimating the estimand of interest. The strategies based on defining ICEs as outcomes result in no missing data due to the ICEs but rely on assumptions about defining the outcome. Strategies where ICEs determine what data are relevant rely in various manners on statistical methods and models, which entail assumptions.

These considerations lead to the road map for handling ICEs that is depicted in Figure 7.1. A primary objective and estimand is chosen, the treatment regimen of interest is defined. ICEs are defined as being consistent with or a deviation from the treatment regimen of interest. In the treatment policy strategy, ICEs are not a deviation from the treatment regimen of interest and can therefore be ignored. An important consequence of the treatment policy strategy is that inference is linked to being randomized to the regimen of interest rather than to the causal effects of a specific intervention.

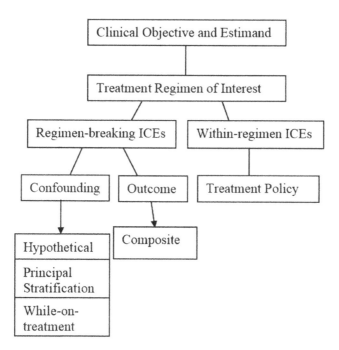

FIGURE 7.1
Conceptual road map for dealing with intercurrent events.

For all strategies other than treatment policy, it is necessary to account for ICEs that are deviations from the defined intervention. The while-on-treatment and composite strategies use definitions to deal with ICEs by including the ICE in some manner as part of the outcome of interest. The hypothetical and principal stratification strategies use statistical methods and models to deal with the confounding caused by the ICEs.

The following chapters provide details and considerations for estimators (analyses) that are consistent with each of the five strategies for dealing with ICEs.

Section III

Estimators and Sensitivity

Section II developed the ideas necessary for choosing and clearly defining estimands in the context of an overall study development process. Section III begins with a brief chapter noting the desired attributes of an estimator in the framework outlined in the ICH E9(R1) Addendum. Subsequent chapters in Section III cover general modeling considerations for longitudinal clinical trial data, along with overviews of specific analytic methods and considerations for each of the five strategies for dealing with intercurrent events that are outlined in the ICH Addendum. Section III closes with chapters dedicated to handling missing data and an overview of sensitivity analyses.

8

Overview of Estimators

The ICH E9(R1) Addendum stipulates the following desired attributes of an estimator: "For a given estimand an aligned analytic approach, or estimator, should yield an estimate as the basis for reliable interpretations. The extent of assumptions is an important consideration for whether an estimate will be robust because the assumptions will often be unverifiable. Therefore, explicitly stating assumptions for primary and sensitivity estimators is essential. Assumptions should be justifiable and implausible assumptions avoided. Sensitivity analyses aligned to the estimator should assess the robustness of results to departures from the underlying assumptions" (ICH, 2019).

The ICH E9(R1) Addendum emphasizes the need to account for intercurrent events (ICEs), which are post-randomization events that may be related to treatment and outcome. Historically, post-ICE data were often treated as missing, and hence analytic considerations focused on the issues arising from incomplete data. Given that some data relevant to an estimand will (almost) always be missing, the issue of missing data has not gone away. However, now missing data is just one part of the broader issue of ICEs. In fact, in this new paradigm, missing data is not the root "problem." ICEs are the problem, one consequence of which can be missing data.

The addendum specifies five general strategies for dealing with ICEs: treatment policy, composite, hypothetical, principal stratification, and while on treatment (ICH, 2019). As noted in Section II, it is important to distinguish ICEs that are part of the intervention under assessment from those that represent deviations from that intervention (Ratitch et al., 2019a). In the treatment policy strategy, ICEs are not a deviation from the regimen of interest and can therefore be ignored in analyses. The other four strategies either define ICEs as an outcome or define what data are relevant.

The strategies based on defining ICEs as outcomes result in no missing data due to the ICEs but rely on assumptions about defining the outcome. Strategies where ICEs determine what data are relevant such that occurrence of ICEs may lead to missing data rely in various manners on statistical methods and models, which entail assumptions. It is possible to explicitly test some assumptions from the data, for example, normality and linearity.

However, it is not possible to test missing data assumptions from the observed data. Sensitivity analyses focus on the nontestable assumptions (Mallinckrodt et al., 2019).

The following chapters provide overviews and considerations for analyses and sensitivity analyses that are consistent with each of the five strategies for dealing with ICEs.

9

Modeling Considerations

9.1 Introduction

Some of the strategies for dealing with intercurrent events (ICEs) condense the longitudinal structure of the clinical trial data into a single observation per patient, while other strategies retain the longitudinal structure. For those strategies that retain the longitudinal structure, certain modeling considerations are important regardless of how ICEs are handled. This chapter provides an overview of modeling consideration for longitudinal data that were covered in detail in Mallinckrodt and Lipkovich (2017, Chapters 6–11), along with considerations for time-to-event analyses and categorical data.

Given the variety of scenarios that may be encountered in longitudinal clinical trials, no universally best model or modeling approach exists. This implies the need to tailor the analysis to the situation at hand. To a certain extent, the study design drives data characteristics, and therefore appropriate analyses follow logically from the design. Given the design is matched to the estimand, the analyst must therefore understand how data characteristics and estimands influence the choice of estimator (analysis).

Common modeling considerations for the analysis of longitudinal clinical trial data include the choice of the dependent variable, how to model mean trends over time, how to account for correlation between repeated measurements, whether covariates should be included, and if so, how best to include them, and how to check or validate models. Additional considerations apply to the modeling of outcomes for which normality and linearity assumptions do not hold, for example, categorical and time-to-event data.

9.2 Longitudinal Analyses

9.2.1 Choice of Dependent Variable and Statistical Test

This section summarizes information in Mallinckrodt and Lipkovich (2017, Chapter 6).

A logical place to begin in developing analyses for longitudinal clinical trial data is to determine the dependent (i.e., response or outcome) variable and the statistical test. These may seem like obvious decisions dictated by the estimand, the design of the trial, and historical precedent. However, several subtleties come into play.

For continuously distributed outcomes, the dependent variable could be the actual outcomes collected, changes from baseline, or percent changes from baseline. Moreover, the statistical test used to evaluate these outcomes could be a contrast at a single, landmark time point, or based on multiple time points, such as a slope analysis, treatment main effect, or treatment-by-time interaction. Convention may guide choices for the primary estimand and primary analysis. Nevertheless, understanding the consequences of the conventional choice is important. It may be necessary for secondary analyses to consider alternatives or to change the customary primary choice when specifying the estimands.

In clinical trials, it is common to base the primary estimand and analysis on a contrast at a landmark time point. Such contrasts are termed cross-sectional contrasts because they focus on a "cross-section in time." However, other "longitudinal contrasts" that involve multiple time points, for example, the entire longitudinal profile, are often worth considering for the primary estimand and analysis or secondarily (supplementary) analyses. An important distinction for cross-sectional contrasts is whether the analysis includes data from just the landmark time point or whether the cross-sectional contrast is derived from the entire longitudinal sequence of data. Incorporating all data even when focus is on a landmark time point opens many possibilities for analyses. An important benefit from longitudinal analyses, regardless of whether focus is on a landmark time point, is in accounting for missing data (see Chapter 15 for further details on accounting for missing data).

Additional benefits from considering more than just a landmark time point include situations with treatment-by-time interactions wherein the difference between treatments varies over time. For example, the two groups may be equal at the landmark time point, but the effect evolved more slowly in one group, thereby yielding transitory differences between treatments that analyses at a landmark time point fail to detect. Another example is when the treatment effect increases over time. Focus may be on the rate of change over time rather than the treatment contrast at endpoint. Alternatively, consider a rapidly evolving and sustained treatment effect. Consistency of the differences between treatments suggests the treatment main effect, which includes data from all time points, could be a useful primary estimand and analysis. Treatment main effects can be more precise, and therefore more powerful than cross-sectional contrasts because they include more data. Interpretation of a treatment main effect is similar to an area under the curve analysis because it is in essence an average effect over all time points.

In situations with more than two treatments, additional options for primary test statistics exist. For example, a global F test at a landmark time point to assess whether any differences exist between treatments at that visit. Alternatively, if treatment main effects are relevant, a global F test for the treatment main effect can be used that includes data from all visits in all arms. Or, a global F test for a treatment-by-time interaction that also includes data from all treatments and all visits can be used.

Clinical relevance is an important factor in choosing how to express continuous outcomes. In some scenarios, the actual outcomes are used as the dependent variable in analyses because they are easy to understand and interpret. However, medical interventions often elicit change in a condition. Therefore, even if the actual outcome is meaningful and easy to understand, change from baseline may be the more intuitive and relevant outcome. For other outcomes, the actual values may not be readily meaningful or widely understood. In these instances, relative changes such as percent change from baseline are easier for broader audiences to appreciate.

However, it is important to consider the analytic consequences of the choices. Perhaps foremost among these analytic consequences is the distribution of the dependent variable. Actual values and percent changes have truncated distributions. Actual values are truncated at the boundaries of the scale range. When assessing symptom severity with scales where zero indicates no symptoms, highly effective treatments will have many patients with scores of 0. Similarly, percent changes are usually truncated at 100%. In contrast, change from baseline is not truncated and tends to be more symmetrically distributed. The importance of this point depends on the analysis. What is the most important for inference is the distribution of residuals after adjusting for covariates and often that distribution is normal or close to normal even though the original measure is skewed.

Considerations for percent change are particularly important when there is no minimum requirement for baseline severity, as would be the case for many secondary outcomes. Some patients may have minimal symptom severity at baseline and small increases or decreases at endpoint. The low baseline severity can result in extreme outlier values for percent change even though absolute changes were small. In situations where percent change is customary, an alternative is to conduct the analyses based on changes from baseline and then convert the mean changes into *percent mean change*, rather than using percent change as the dependent variable to assess *mean percent change*.

9.2.2 Modeling Covariance (Correlation)

This section summarizes information in Mallinckrodt and Lipkovich (2017, Chapter 7).

In longitudinal clinical trial data, correlations between repeated measurements *within* the same patients often arise from between-patient variability and from adjacency in time or location of evaluations. Therefore, assuming independence of observations is seldom justified. Failure to account for the association between observations can result in biased point estimates and standard errors that do not properly reflect the uncertainty in the data.

The association between repeated measurements on the same patients can be modeled as a function of the random effects, the residual effects, or both. Random measurement errors contribute to variability and are a modeling consideration. In longitudinal settings, it is particularly important to consider the potential for error variance to increase or decrease over time (Mallinckrodt et al., 2003).

In clinical trials, the random effects are seldom of direct interest because focus is primarily on the fixed effects. Therefore, need rarely exists to explicitly model the random effects, and they can therefore be modeled as part of the within-patient error (residual) correlation structure. In such cases, the patient-specific effects and serial correlations combine, with or without changes in error variance over time, to yield what is often an unstructured correlation pattern for the within-patient (residual) errors. Care must be taken when there are multiple repeated within-patient factors (e.g., repeated measurements over time for the two eyes on the same patient). In this example, both time and eye are repeated factors and need to be accounted for when modeling within patient covariances.

In clinical trials, the number of assessment times relative to the number of patients is often small. Such situations are amenable to unstructured modeling of within-patient covariance (residual errors) that also accounts for the patient-specific random effects. Unstructured modeling places few restrictions (assumptions) on the model and is often a preferred approach, especially in large, confirmatory trials.

The number of parameters for unstructured modeling of correlations increases by $n(n+1)/2$, where n is the number of assessment times. However, in many clinical trial settings, it is unlikely that the number of covariance parameters to estimate is prohibitively large, unless separate unstructured matrices are fit for multiple groups (e.g., treatment arms) within the same data set.

Model fitting criteria can guide the choice of the best fitting model after data becomes available. Alternatively, an unstructured (or other appropriate) approach can be prespecified and alternative structures tested for better fit, or in case of failure to converge. See Mallinckrodt and Lipkovich (2017, Chapter 7), for a detailed discussion and examples of fitting various correlation structures.

9.2.3 Modeling Means Over Time

This section summarizes information in Mallinckrodt and Lipkovich (2017, Chapter 8).

In longitudinal clinical trials, means can vary over time due to, among other things, study effects and the natural evolution of the disease. Even when focus is on a single landmark time point (e.g., endpoint visit), properly understanding and modeling the time trends are important.

In mixed-effects models, time can be modeled in either a structured or unstructured manner. In unstructured modeling of time, the time variable is a categorical (i.e., class) effect. The fixed-effect solutions in such a mixed model represent the unique effects of each assessment time and have $t - 1$ degrees of freedom for t time points. In structured modeling of time, the time variable is a continuous independent variable (covariate) as is typically done in regression and covariate analyses. Therefore, structured models for time can use fewer degrees of freedom and be more powerful than unstructured approaches if the functional form of the time trend is correctly specified. In structured approaches, the time variable can be considered strictly a fixed effect, strictly a random effect, or have both fixed and random components. Models with time as a random effect are termed random coefficient regression models.

Although the longitudinal pattern of treatment effects is usually of interest, the functional form of the mean responses over time may be difficult to anticipate during protocol development. Linear time trends may not adequately describe the mean responses. Nonlinear trends may arise from inherent characteristics of the disease and the drug under study and/or from trial design features. For example, if titration dosing is used with initial dosing at a subtherapeutic level to reduce adverse events, there may be a lag period with little or no improvement. Conversely, if a drug has rapid onset of a fully therapeutic effect, the beneficial effects may increase rapidly across early assessments and then level off thereafter. In such cases, parsimonious approaches to modeling means over time may lead to inaccurate results, and more general unstructured models may be preferred.

Therefore, in many scenarios, an unstructured modeling of means over assessment times requires fewer assumptions, does not require estimation of an inordinate number of parameters, and is dependable in yielding a useful result (Mallinckrodt et al., 2003). Or, several ways of parameterization can be used to fit segmented or piecewise trajectories, sometimes referred to as broken stick or hockey stick models.

Another consideration in modeling mean trends over time regards other covariates that are included in the analysis and their interaction with time. For example, the effect of baseline severity may be an important covariate because patient responses may depend on their condition at the start of the trial. It is usually preferable to allow a full interaction of covariates with time because not doing so imposes a restriction that the dependence of response on the covariate is the same at all assessment times (Mallinckrodt et al., 2008, 2014). As with fitting covariance structures, care must be taken with doubly repeated measurements to fit a proper mean structure. See Chapter 8 in Mallinckrodt and Lipkovich (2017) for detailed discussion and examples of SAS and R code for how to implement various approaches to modeling mean trends over time.

9.2.4 Accounting for Covariates

This section summarizes information in Mallinckrodt and Lipkovich (2017, Chapter 9).

Covariates are often included in longitudinal analyses to account for the variability due to these effects, thereby increasing precision in estimates of treatment effects. Fitting covariates can also counteract bias if values of continuous covariates or levels of categorical covariates are not balanced across treatments. Previously, it was noted that it is usually preferable, or at least it should be considered, to allow a full interaction of covariates with time. Fitting covariate-by-time interactions avoids imposing restrictions that the dependence of response on the covariates is the same at all assessment times.

Covariate-by-treatment interactions, commonly called subgroup analyses when the covariate is a categorical effect and subclass regression when the covariate is a continuous effect, are included in longitudinal analyses to assess the consistency of treatment effects across levels of the covariate. This is typically done in supportive or exploratory analyses. Consistency of the treatment effect across subgroups indicates that the average treatment effect is in general applicable regardless of the specific characteristic described by the covariate. Substantial heterogeneity in treatment effect may indicate that treatment benefit pertains to only a subset of the population. However, apparent heterogeneity in the observed treatment effect across subgroups can arise due to chance when partitioning the sample into several subgroups; this chance increases as the number of subgroup factors increases. Furthermore, clinical trials are generally not powered to detect heterogeneity in treatment effects; thus statistical tests may fail to detect heterogeneity due to low power (Alosh et al., 2015). For more details on model building, see Verbeke and Molenberghs (2000). For detailed discussion on various aspects of fitting covariates in longitudinal models, including examples of SAS and R code, see Mallinckrodt and Lipkovich (2017, Chapter 9).

Fitting baseline severity as a covariate versus fitting it as part of the response vector is a particularly interesting consideration. In an analysis of covariance (ANCOVA) model, baseline is a fixed-effect covariate. Although fitting baseline severity as a covariate is a common approach, it is not the only option. Baseline severity is a special case of covariate adjustment because baseline severity can be a response at time 0 and therefore may be accounted for by including baseline outcomes as part of the response vector (Liang and Zeger, 1986). Two methods that include baseline as a response are the so-called LDA (longitudinal data analysis) and cLDA (constrained longitudinal data analysis) (Liu et al., 2009). The constraint in cLDA is that the baseline values are equal in the intervention groups.

In the LDA and cLDA models where baseline is a response, it is a random effect. The ANCOVA model tends to be the easier and the more common approach. However, when an appreciable number of baseline observations are missing, or when an appreciable number of patients have a baseline but no post-baseline values, the LDA or cLDA approaches can have advantages. See Mallinckrodt and Lipkovich (2017, Chapter 9) for detailed discussion and examples of implementing the ANCOVA, LDA, and cLDA approaches in SAS and R. The handling of baseline covariates is also discussed in Verbeke and Molenberghs (2000, Chapter 13).

9.2.5 Categorical Data

This section summarizes information in Mallinckrodt and Lipkovich (2017, Chapter 10).

For a comprehensive review of categorical data analyses, see Agresti (2002). McCullagh and Nelder (1989) provide specific focus on generalized linear models, which are used extensively in categorical data analyses. Many of the principles guiding analyses of continuous outcomes also apply to categorical outcomes. For example, considerations regarding modeling means over time and correlations between repeated measurements are essentially the same as previously outlined for continuous outcomes. Additional similarities include the ability to formulate a longitudinal categorical analysis with treatment contrasts that exactly match the results from an analysis of the endpoint time only. However, analyses of categorical endpoints entail additional complexity compared with continuous endpoints that are typically modeled using linear models with Gaussian error distributions. This additional complexity stems from the (typically) nonlinear relationship between the dependent variable and independent variables, and from the non-normal distribution of errors. This is further exacerbated by the so-called mean-variance relationship, whereas for Gaussian data, whether univariate or hierarchical, the mean and variance-covariance functions are functionally independent.

Conceptually, accommodating these aspects is as simple as specifying an appropriate link function to account for nonlinearity and an appropriate distribution for the errors. However, these accommodations necessitate additional computational complexity and intensity, which in turn limits flexibility in implementing analyses. Moreover, unlike the "linear normal" modeling of continuous outcomes where a single result has both a marginal and a hierarchical (random effects) interpretation, the parameters estimated from marginal and random-effects models with categorical data, and more broadly for any generalized linear mixed model with a nonidentity link function, describe different effects and have different interpretations (Jansen et al., 2006). Therefore, scientific goals should drive the choice between marginal and random-effect models, but the computational complexity and availability of software tools are also considerations.

Conceptually, the distinction between marginal and random-effects models is that marginal models estimate the average response in the population, whereas random-effects models estimate responses for the average patient (the patient for which the random effect = 0). In linear mixed-effects models with normally distributed data, the average response and the response of the average patient are the same because the random effects cancel out when taking expectation over the outcome variable. This is due to the additivity of linear models with an identity link and random effects having a mean and expected value = 0. Hence, fixed-effect parameters estimated from a linear mixed-effects model in Gaussian data have both a marginal and hierarchical interpretation. In non-normal data, these two targets of inference are not identical because the function connecting outcome variables and covariates is nonlinear. In regulatory settings, population average effects are typically the focus.

A similar consideration applies regarding covariates. When a covariate is included in the model, estimates and inferences are conditional on the covariate. By default, analyses adjust to the mean value of continuous covariates. When the association between the response and the covariate is linear, as in Gaussian data, the population average response and the response of a patient with the mean covariate value are the same. Therefore, in Gaussian data including versus not including covariates does not influence the target of inference. However, in non-Gaussian data including versus not including the covariate does influence the target of inference because the expected value of the population average is not equal to the expected value for the patient with the average covariate value. Mallinckrodt and Lipkovich (2017, Chapter 10) provide technical details and example code in SAS and R to illustrate these points.

In a generalized linear model (GLM), the log-likelihood is well defined, and an objective function for estimation of the parameters is straightforward to construct. In a generalized linear mixed-effects model (GLMM), categorical data pose increased complexity over continuous data. These complexities hinder construction of objective functions. Even if the objective function is feasible mathematically, it still can be out of reach computationally (SAS, 2013). These complexities and restrictions have led to the development of several estimation methods that typically either approximate the objective function or approximate the model, for example, using Taylor series expansions.

Generalized estimating equations (GEE; Liang and Zeger, 1986; Fitzmaurice et al., 2004; Molenberghs and Verbeke, 2005) can circumvent the computational complexity of likelihood-based analyses of categorical data and is therefore a viable alternative whenever interest is restricted to the mean parameters (treatment difference, time evolutions, effect of baseline covariates, etc.). GEEs are rooted in the quasi-likelihood ideas expressed by McCullagh and Nelder (1989). Modeling is restricted to the correct specification of the marginal mean function, together with so-called working assumptions about the correlation structure among the repeated measures.

However, GEE is valid only when data are missing completely at random (MCAR, see Chapter 15). Therefore, in the frequent situations where missing data is concerned, GEE is often implemented with inverse probability weighting to yield weighted generalized estimating equations (wGEE) to ascertain validity under MAR (see Chapter 20 for additional details and examples of implementing wGEE in SAS and R). Alternatively, GEE can be combined with multiple imputation as another way of achieving validity under MAR.

9.2.6 Model Checking and Verification

This section summarizes information in Mallinckrodt and Lipkovich (2017, Chapter 11).

With likelihood-based estimation, the objective function can be used to assess model fit and to compare the fit of candidate models. If one model is a submodel of another, likelihood ratio tests can be used to assess whether the more general model provides superior fit. The likelihood ratio test is constructed by computing −2 times the difference between the log likelihoods of the candidate models. This statistic is compared to the χ^2 distribution with degrees of freedom equal to the difference in the number of parameters for the two models.

For models that are not submodels of each other, various information criteria can be used to determine which model provides the best fit. The Akaike information criteria (AIC) and Bayesian information criteria (BIC) are two common approaches (Hurvich and Tsai, 1989). These criteria are essentially log-likelihood values adjusted (penalized) for the number of parameters estimated and can thus be used to compare different models. The criteria are set up such that the model with the smaller value is preferred.

Important assumptions required for valid regression-type analyses of continuous outcomes include linearity, normality, and independence (Wonnacott and Wonnacott, 1981). The first assumption is about the true form of the mean function, and the last two assumptions are about the error term. Errors (i.e., residuals) are the difference between observed and predicted values. Because the assumption of independence is seldom justifiable, longitudinal analyses include covariances between the repeated measures.

An inherent challenge in model checking – evaluating the mean structure and the errors – is that they are conditional on one another. Therefore, checking assumptions is typically an iterative process that entails refitting the model. For example, the errors may have some type of skewed distribution due to an incorrect mean model. After altering the mean model, the residuals can be rechecked.

Checking for patterns of residuals, such as associations between residuals and covariates, is also needed to ascertain whether the mean function is appropriately specified; that is, are all the important effects fitted and is the modeled form of the association (linear, etc.) correct? This again emphasizes the iterative nature of model checking. After identifying such patterns, the model is changed, and the residuals reevaluated.

Outliers are anomalous values in the data that may have a strong influence on the fitted model, resulting in a worse fit for most of the data. The important issue to address is whether outliers had an important effect on results, not whether outliers or non-normality existed.

Another important aspect of model checking is to assess the influence of clusters of observations on model fit. In clinical trials, the clusters of interest usually include patient and investigative site. The general idea of quantifying the influence of clusters relies on computing parameter estimates based on all data points, removing the cases in question from the data, refitting the model, and computing statistics based on the change between the full-data and reduced-data estimations. Again, the main idea is not so much whether influential clusters existed but rather to understand what impact the most influential clusters had on results. Mallinckrodt and Lipkovich (2017, Chapter 11) provide additional details and examples of influence and residual diagnostics, including SAS code.

9.3 Summary

Regardless of the estimand and how ICEs are handled, certain modeling principles apply for analyses of longitudinal data. Key considerations include:

- Actual change from baseline will usually have better distributional properties than percent changes, especially when there is no baseline minimum requirement for the dependent variable.

- The association between repeated measurements on the same patients can be modeled as a function of the random effects, the residual effects, or both. No approach is universally better than another. The analysis must be tailored to the situation at hand. However, unstructured modeling of the residual effects is often a preferred strategy, especially when the number of observations at each assessment time is not small.

- Unstructured modeling of mean trends over time is possible because the number of parameters relative to the number of observations is not large. This is important, especially in confirmatory studies, because an unstructured model for the means requires no assumption about the time trend and models with fewer assumptions are preferred. However, more parsimonious models can be more powerful.

- In ANCOVA, models where baseline is accounted for as a covariate, baseline is considered a fixed effect. In LDA and cLDA models where baseline is considered a response, it is a random effect. The ANCOVA model tends to be the easier and the more common approach. However, when an appreciable number of baseline observations are missing, or when an appreciable number of patients have a baseline but no post-baseline values, the LDA or cLDA approaches can have advantages.

10

Overview of Analyses for Composite Intercurrent Event Strategies

10.1 Introduction

In composite strategies, the presence or absence of relevant intercurrent events (ICEs) is a component of the outcome variable (ICH, 2019). Composite strategies are flexible and therefore adaptable to a variety of scenarios. A key aspect of composite strategies is that the outcome definition results in no missing data due to ICEs, although some data may still be missing due to other reasons (Mallinckrodt et al., 2019).

In some composite strategies, the presence of an ICE is the most meaningful outcome. For example, in many clinical trial scenarios, death is an ICE; however, in clinical trials that assess rates of myocardial infarction, death may be more meaningful than observations before death, the observations after death will not exist, and it may not be possible to ascertain whether a patient who died had, or would have had, a myocardial infarction. Using a primary endpoint that is a composite of death or myocardial infarction resolves this issue (ICH, 2019).

Other examples of composite strategies include defining an outcome variable based on the combination of the presence or absence of relevant ICEs and the observed clinical outcome. For example, patients can be assigned an unfavorable outcome if they have a relevant ICE; for patients who did not have a relevant ICE, the observed clinical assessments determine the outcome.

10.2 Ad Hoc Approaches

Historical examples of analyses consistent with composite strategies include non-responder imputation (NRI) for binary endpoints and baseline observation carried forward (BOCF) for continuous endpoints (Mallinckrodt and

Lipkovich, 2017; Mallinckrodt et al., 2017). In NRI, patients are considered a treatment success if they complete the treatment period on their randomized treatment regimen and achieve a certain level of improvement. Patients are considered a treatment failure if they do not achieve the level of improvement or, regardless of improvement, discontinue the study or study treatment early.

As such, the acronym "NRI" is a misnomer because there is no explicit imputation. Every patient has an observed outcome for treatment success/ failure. This results in complete, binary data for each patient. The population parameter to be estimated and compared between treatments is the proportion of treatment successes, which can be estimated via logistic regression or related methods suitable for binary data (Mallinckrodt et al., 2019).

The acronym "BOCF" implies nonstochastic imputation for longitudinal data. However, BOCF as commonly implemented is consistent with composite strategies for dealing with ICEs (Mallinckrodt and Lipkovich, 2017; Mallinckrodt et al., 2017).

In BOCF, data prior to relevant ICEs are ignored, and the baseline value of the outcome variable is used as the outcome variable for patients with ICEs. The observed endpoint outcomes are used for patients who do not have relevant ICEs. Using the baseline value as the outcome for patients with ICEs yields zero change from baseline – no benefit – an unfavorable outcome. The population parameter to be estimated and compared between treatments is the mean at a landmark (endpoint) time point, which is commonly done with analysis of variance (ANOVA) or analysis of covariance (ANCOVA), typically ignoring the repeated measure nature of the data (Mallinckrodt and Lipkovich, 2017; Mallinckrodt et al., 2017).

Modified approaches to NRI and BOCF (mNRI and mBOCF) prespecify a set of treatment-related reasons for early discontinuation, such as lack of efficacy or intolerability; NRI or BOCF is applied to patients with these ICEs. However, discontinuations for reasons unrelated to study treatment, such as having relocated and no longer being able to attend study visits, are not relevant ICEs, and the observed clinical outcomes are used in some manner to estimate what would have been observed if the patients had not discontinued (Mallinckrodt et al., 2019).

Both NRI and BOCF yield complete data and hence no assumptions about missing data are required. Although it may be reasonable to assume there is no lasting benefit from drug after discontinuation/rescue in trials of symptomatic interventions, defining no lasting benefit as zero change for BOCF and non-response in NRI ignores changes from nonpharmacologic sources such as study effects and placebo effects (O'Neill and Temple, 2012). Therefore, BOCF and NRI are unbiased estimators of the intended estimand only when there is zero mean change in a placebo group (BOCF) or the response rate for placebo is zero (NRI) (Mallinckrodt et al., 2012; Mallinckrodt et al., 2013). Reference-based imputation (see Chapter 22) provides more principled alternatives to BOCF and NRI; therefore, use of BOCF and NRI is not justified in most situations (Mallinckrodt et al., 2019).

10.3 Rank-Based and Related Methods

Many nonparametric methods are based on the ranks of the observations instead of the actual values. Commonly used nonparametric methods to assess ranks in clinical trials include the Wilcoxon signed-rank test and Mann–Whitney test (Shih, 2002). Deviations from the treatment regimen under assessment (ICEs) can be accounted for in these methods by ranking the events according to their severity (Mallinckrodt et al., 2019). In the case of early-study discontinuation, the rankings could be based on the reasons for withdrawals (Gould, 1980) and the time of withdrawal (Lachin, 1999). For example, death may receive the worst rank, followed by lack of efficacy, then adverse reactions, and patient refusal. Within the same category of withdrawal, early dropouts could be given worse ranks than later dropouts. After assigning ranks, the usual testing procedure is conducted (Shih, 2002).

Rank-based methods entail several important considerations. Unless estimation targets quantiles (e.g., medians), rank-based methods do not estimate the treatment effect in the original measurement unit. Therefore, it can be difficult to interpret the clinical relevance of differences between treatments. Moreover, it may be difficult to assign proper rankings. For example, is death the worst outcome if it was unrelated to treatment? Is discontinuation for lack of efficacy worse than discontinuation for an adverse event? This decision rests on the severity and duration of the adverse event relative to the degree of improvement or worsening that resulted in early discontinuation for lack of efficacy. Therefore, in many situations, assumptions about the rankings or adjudication of each ICE are required (Mallinckrodt et al., 2019).

The win ratio approach overcomes some of the difficulty in interpreting differences in ranks (Wang and Pocock, 2016). Patients in the experimental and control treatments are paired in all possible combinations. Patients in the experimental treatment group are labeled as winner (loser) if they have the more (less) favorable outcome in each pair of patients. The population parameter estimated is the win ratio, which is the total number of winners divided by the total number of losers. Confidence intervals can be obtained using the bootstrap (Wang and Pocock, 2016) or other means (Kawaguchi and Koch, 2015). However, this method is not recommended when the number of ties is appreciable (Wang and Pocock, 2016), and it does not overcome the difficulties in assigning ranks. A similar alternative is the proportion in favor of treatment approach, which is based on the proportion of pairwise comparisons that are in favor of treatment over control.

The trimmed mean approach deals with early discontinuation in a manner that is intended to overcome some of the problems of interpretability in ranks but does so at the cost of changing the target population. The approach is applicable when comparing treatments intended to improve the signs and symptoms of a disease rather than alter the underlying disease mechanism (Permutt and Li, 2017). Mehrotra et al. (2017) provided additional examples, and Wang et al. (2018) evaluated the approach with an extensive simulation study.

The basic idea is to assign an arbitrarily bad outcome to each patient that discontinued early (or more generally, had a relevant ICE) such that the assigned outcomes are worse than any observed outcomes. An equal fraction of the worst outcomes is trimmed from each treatment group such that all patients with relevant ICEs are trimmed, thereby obviating concerns about sensitivity to the actual score assigned and yielding an analysis data set with no missing values. Means (medians) or mean (median) changes from the trimmed values are calculated, and permutation methods are used to construct a reference distribution for hypothesis testing and confidence intervals (Permutt and Li, 2017).

The specific estimand addressed by the trimmed mean is the difference between treatments in endpoint means in the X% subset of patients with the most favorable outcomes, where patients with relevant ICEs have an outcome worse than any patient in the X% subset. The percent of data to be trimmed (100-X) is determined using either an a priori chosen fixed percentage or adaptive trimming based on the actual results of the trial (Permutt and Li, 2017). It is not clear that the distributional properties of trimmed means apply with the data-dependent adaptive trimming. Moreover, adaptive trimming leads to estimands defined by the data and therefore should be avoided in favor of fixed trimming (Mallinckrodt et al., 2019). However, even with fixed trimming, the target population is not well defined, either in terms of baseline patient characteristics or what level of outcome corresponds to the best X% of patients.

Trimming favors the group with fewer dropouts (or more generally, fewer ICEs) because having more completers can be a beneficial effect of the drug, or conversely, higher dropout can be a bad effect. In a simulation study, trimmed means yielded significant differences between treatments in situations where there was no difference between treatments in the corresponding untrimmed data set (Wang et al., 2018). Therefore, the utility of the trimmed mean hinges on the reasonableness of its assumptions and on the ability to relate conclusions to specific patient populations. These assumptions include that dropout (or more generally any relevant ICE) is an equally bad outcome in all patients (Wang et al., 2018). Permutt and Li (2017) noted scenarios where this assumption is not valid, such as dropout due to intolerability may not be as bad as dropout due to death. These authors also noted that increased adherence is not always beneficial if that adherence arises from pleasant side effects of the drug that is unrelated to the outcome.

Another assumption, as with all methods that incorporate discontinuation or rescue (or other ICEs) as part of the outcome, is that adherence/

rescue decisions in the clinical trial are sufficiently close to the decisions in clinical practice for the results to be generalized. This assumption deserves scrutiny in placebo-controlled and blinded trials. Placebo control and blinding are not part of general clinical practice, and they influence outcomes (Mallinckrodt, Molenberghs, and Rathmann, 2017; Mallinckrodt and Lipkovich, 2017).

Power is another important consideration. The reduction in power from using a subset of the data could be considerable compared to other composite strategies. Therefore, study planning should include the impact of trimming (Wang et al., 2018). Although previous studies on the same compound in similar settings can be a useful guide, uncertainty in rates of dropout and other ICEs may add additional uncertainty into sample size estimation (Mallinckrodt et al., 2019).

It is also important to consider that the trimmed mean assesses a unique estimand that may not be relevant in all situations. For example, trimmed means estimate benefit in a subset of patients. However, medications have cost for everyone who takes the drug. Therefore, trimmed means may not be relevant to health technology assessors. Similarly, for those wishing to make across study comparisons, the lack of historical use of the trimmed mean could be problematic (Wang et al., 2018). Moreover, trimming is not consistent with the intention-to-treat approach because the population defined in the estimand is a subset of all randomized patients and that subset is not well defined (Mallinckrodt et al., 2019).

10.4 Summary

In composite strategies, the presence or absence of relevant ICEs is a component of the outcome variable. Composite strategies are flexible and therefore adaptable to a variety of scenarios. A key aspect of composite strategies is that the outcome definition results in no missing data due to ICEs. BOCF and NRI are unbiased estimators of the intended estimand only when there is zero mean change in a placebo group (BOCF), or the response rate for placebo is zero. Therefore, the use of BOCF and NRI may not be justified in many situations.

Rank-based methods can be used in composite strategies by assigning ranks to outcomes for patients with ICEs that are worse than ranks for observed outcomes. Results from rank-based methods can be difficult to interpret because treatment effects are not estimated in the original measurement units. Moreover, assigning ranks typically requires either assumptions about the relative severity of various ICEs or adjudication. In the win ratio approach, patients in the experimental and control treatments are paired in all possible combinations. Patients in the experimental group are labeled as winner if they have the more favorable outcome. The population parameter estimated is the win ratio, the total number of winners divided by the total number of losers; in other words, the odds of doing better versus worse on experimental treatment compared to control.

The trimmed mean attempts to overcome the problems of interpretability from losing the original unit of measure when assigning ranks. It is applicable when comparing treatments on signs and symptoms of a disease rather than altering the underlying disease mechanism. The approach assigns arbitrarily bad outcomes to each patient with an ICE such that the assigned outcomes are worse than any observed outcomes. An equal fraction of the worst outcomes is trimmed from each treatment group such that all patients with relevant ICEs are trimmed. Means (medians) or mean (median) changes from the trimmed values are calculated and permutation methods used for hypothesis testing and confidence intervals. The estimand addressed by the trimmed mean is the difference between treatments in endpoint means in the X% subset of patients with the most favorable outcomes, where patients with relevant ICEs have an outcome worse than any patient in the X% subset. However, this X% subset is not a well-characterized target population. Assumptions in the trimmed mean approach include that all ICEs handled by the composite strategy are an equally bad outcome in all patients, and as with all methods that incorporate discontinuation or rescue (or other ICEs) as part of the outcome is that adherence/rescue decisions in the clinical trial are sufficiently close to the decisions in clinical practice for the results to be generalized.

11

Overview of Analyses for Hypothetical Intercurrent Event Strategies

11.1 Introduction

Hypothetical strategies for dealing with intercurrent events (ICEs) are useful when a reasonable scientific question stems from what outcomes would have been under some hypothetical (counterfactual) condition(s). For example, when rescue medication is ethically necessary, a treatment effect of interest might be the outcomes if rescue had not been taken. Care is required to describe precisely the hypothetical conditions reflecting the scientific question of interest in the context of the specific trial (ICH, 2019).

In principle, the number of hypothetical scenarios can be large. However, three general categories of hypothetical conditions reflect many potential applications: (1) Outcomes for the initially randomized treatments if relevant ICEs had not occurred. (2) Outcomes for the treatment regimens if relevant ICEs had not occurred. (3) Outcomes for a treatment regimen that was not included in the trial or that was included but patients did not follow. Scenarios 1 and 2 involve what would have happened if patients had adhered to their intervention. The only difference between scenarios 1 and 2 is whether use of rescue medication is part of the treatment regimen (scenario 2) or a departure from it (scenario 1) (Mallinckrodt et al., 2019). Therefore, the same analytic approach can be applied to both scenarios 1 and 2.

In scenario 1, the target of inference is the initially randomized treatments, drug vs. control, and rescue is an ICE – a departure from the treatment under investigation. In scenario 2, the target of inference is drug plus rescue vs. control plus rescue, and therefore rescue is not a departure from the regimen of interest. In scenario 3, the intent is to assess what would have happened if patients took a reference treatment rather than discontinuing medication and/or discontinuing from the trial. If the reference group is placebo, the hypothetical scenario is assessing the effects of a treatment regimen that includes the randomized treatment followed by placebo if patients discontinue the initial medication. This allows estimation of the effects of the initially randomized treatment if treatment had no further benefit in patients that discontinued study medication and, at the same time, accounts for the natural disease trajectory and trial effect. A hypothetical strategy is needed in this situation because it is usually unethical or impractical to expose patients to placebo after discontinuation of an initial medication. This approach is a more principled and preferred alternative to composite strategies such as baseline observation carried forward (BOCF) and non-responder imputation (NRI) that ascribe no benefit to patients who do not adhere to the treatment of interest and do not account for the natural evolution and variability of patients' outcomes over time. If the reference group is an active medication, the hypothetical scenario is assessing a treatment regimen that includes assignment to the initial treatment plus rescue (Mallinckrodt et al., 2019).

11.2 Estimators for What Would Have Happened in the Absence of Relevant Intercurrent Events

11.2.1 Introduction

Many options exist for estimating what would have happened in the absence of relevant ICEs. Typically, data after relevant ICEs are excluded from analyses (considered missing) either because they were not observed or the data that were observed are not relevant in estimating the estimand of interest. As such, methods commonly used in missing data situations are applicable. The choice of specific method depends on plausibility of assumptions. Therefore, missing data assumptions are central in choosing methods for hypothetical strategies. The distinction between missing at random (MAR) and missing not at random (MNAR) is particularly important (Mallinckrodt et al., 2019).

In the well-known taxonomy of missing data mechanisms (Little and Rubin, 2002), data are MAR if, conditional upon the covariates in the analysis and the observed outcomes of the variable being analyzed, the probability of missingness does not depend on the unobserved outcomes (either missing or counterfactual) of the variable being analyzed. Another way to conceptualize MAR is that conditional on the covariates and observed outcomes, the predicted statistical behavior of the unobserved (or counterfactual) data is what it would have been if observed.

Data are MNAR if, conditional upon the covariates in the analysis and the observed outcomes of the variable being analyzed, the probability of missingness depends furthermore on the unobserved outcomes of the variable being analyzed. Another way to conceptualize MNAR is that conditional on the covariates and observed outcomes, the predicted statistical behavior of the unobserved (or counterfactual) data is not what it would have been if observed. In other words, there is something fundamentally different about the unobserved data not foreseeable from the predictions based on observed outcomes and covariates.

Methods valid under MAR include maximum likelihood (ML), restricted ML, multiple imputation (MI), and inverse probability weighting (IPW) (Molenberghs and Kenward, 2007). Some MNAR methods include the controlled-imputation family of methods, selection models, shared parameter models, and a variety of pattern-mixture models (Verbeke and Molenberghs, 2000; Molenberghs and Kenward, 2007; Mallinckrodt and Lipkovich, 2017).

Although no universally best approach exists, when modeling what would have happened if ICEs had not occurred, it is often reasonable to begin with approaches based on MAR and then use MNAR methods to assess sensitivity to departures from MAR (Verbeke and Molenberghs, 2000; Molenberghs and Kenward, 2007; Mallinckrodt and Lipkovich, 2017). When modeling hypothetical scenarios that involve changes in treatment, MNAR methods, particularly controlled-imputation approaches, are useful. Controlled-imputation methods are specific versions of pattern-mixture models that are gaining interest and application because their assumptions are transparent and easy to understand, thereby making it straightforward to match a specific analysis to specific hypothetical scenarios (Carpenter, Roger, and Kenward, 2013; Ratitch, O'Kelly, and Tosiello, 2013; Mallinckrodt and Lipkovich, 2017).

11.2.2 Likelihood-Based Analyses

Foundational references for ML estimation are Rubin (1976), Harville (1977), Jennrich and Schluchter (1986), and Little and Rubin (1987). More contemporary references for practical implementations of ML in clinical trial contexts include Verbeke and Molenberghs (2000), Molenberghs and Kenward (2007), and Mallinckrodt and Lipkovich (2017). It is possible to ignore the missing data process in likelihood-based analyses and obtain unbiased estimates of what would have been observed – if missing data arise from an MAR mechanism (or a missing completely at random [MCAR] mechanism), and the model is correctly specified.

When the aim is to estimate what would have happened in the absence of early discontinuation of the randomized treatment, it is useful to fit a suitable likelihood-based repeated measures analysis to the data observed until the point of early termination. When the aim is to estimate what would have happened if patients had remained on the initially randomized medication and not taken rescue, counter to the fact that rescue was taken, it is relevant to fit a suitable model to the data observed until either the initiation of rescue or early termination (Mallinckrodt and Lipkovich; 2017; Mallinckrodt et al., 2019).

Little and Rubin (2002) provide a proof of why the missing data process can be ignored in likelihood-based analyses if the missingness arises from an MCAR or MAR mechanism. In that proof, the hypothetical "full" data are split into two parts: the complete outcome and missing data indicator, which can also be described as the measurement process and the missingness process, respectively (Verbeke and Molenberghs, 2000). The complete outcome data are further split into observed and missing parts. The full data likelihood is then factored into parts that isolate the missingness parameters from the parameters underlying the outcome process (see Chapter 18 for additional details). Using appropriate link functions and error distributions, ML can be applied to outcomes with a variety of distributions, including normally distributed, categorical, and time to event (Verbeke and Molenberghs, 2000; Molenberghs and Kenward, 2007; see Chapter 18 for additional details on likelihood-based analyses).

11.2.3 Multiple Imputation

MI is another popular method based on MAR. Foundational references for MI include Rubin (1978a, 1987) and Little and Rubin (2002), with contemporary references from Carpenter and Kenward (2013), and van Buuren (2007, 2018), with extensive illustrations of clinical trial applications in Molenberghs and Kenward (2007), O'Kelly and Ratitch (2014), and Mallinckrodt and Lipkovich (2017).

MI has several variations, but the key idea is to impute the missing values using a likelihood-based model, combine the observed and imputed values to create a completed data set, analyze the now complete data using methods appropriate for complete data, and repeat the process multiple times. Results are combined across the multiple data sets for final inference. Uncertainty due to imputation is accounted for by combining within- and between-imputation variance. These steps can be applied to outcomes with a variety of distributions, including normally distributed, categorical, count, and time to event (e.g., with piecewise exponential covariate-adjusted hazard) outcomes (see Chapter 19 for details on MI).

If MI is implemented using the same imputation and analysis model, and that model is the same as the analysis model used in an ML analysis, MI and ML will yield asymptotically similar point estimates, but ML will be somewhat more efficient (Molenberghs and Kenward, 2007; Mallinckrodt and Lipkovich, 2017). However, with distinct steps for imputation and analysis, MI has more flexibility than ML. This flexibility is useful when baseline covariates are missing, when errors are not normally distributed such that likelihood-based analyses are difficult to implement, and when inclusive modeling strategies are used to help account for missing data. MI is also useful for sensitivity analyses (Carpenter and Kenward, 2013; O'Kelly and Ratitch, 2014; van Buuren, 2007, 2018; Mallinckrodt and Lipkovich, 2017).

Intermittent missing data usually do not arise from relevant ICEs, and the MAR assumption is usually reasonable; therefore, standard MI approaches are applicable. For intermittent missing data, Markov chain Monte Carlo (MCMC) sampling is used. Or, when missingness is mostly monotone, but some values are intermittently missing, MCMC can be used to impute the intermittent missing data, and then the easier-to-implement sequential regression approach can be applied to the remaining monotone missing values (Mallinckrodt and Lipkovich, 2017).

For mixed outcome types (continuous and categorical), MI can be implemented either by sequential imputation (for monotone patterns of missing data) or in the case of arbitrary missingness by repeatedly sampling from full conditional distributions (van Buuren, 2007). In fully conditional specification (FCS), unlike joint modeling, the multivariate distributions are specified through a sequence of conditional densities of each variable given the other variables. With FCS, it is possible to specify models for which no joint distribution exists – a situation referred to as incompatibility. Although incompatibility is not desirable, in practice it is a relatively minor problem, especially if the missing data rate is modest (van Buuren, 2018).

11.2.4 Inverse Probability Weighting

IPW is another MAR approach. A foundational reference for IPW is Robins, Rotnitzky, and Zhao (1995), with clinical trial applications illustrated in Molenberghs and Kenward (2007), and Mallinckrodt and Lipkovich (2017). IPW uses a model to estimate the probability of dropout (or more generally any ICE) given the observed data. The inverse of these probabilities is applied to the observed data to create a pseudosample reflecting what would have been observed if no data were missing. The motivation behind IPW is to correct for bias caused by nonrandom selection (see Chapter 20 for technical details on IPW).

An important requirement of IPW is that all possible values of the outcome variable have nonzero probability of being observed. If some values have low probabilities, it is unlikely that these values would be observed in a sample. IPW may lead to bias if some values are not observed in the sample and/or to high variance if some values are observed but have low estimated probabilities resulting in large weights.

Weighting can be at the patient level, with one weight per patient reflecting the inverse probability of observing the dropout pattern, or weighting can be at the observation level, with one weight per patient per visit that reflects the inverse probability of a patient remaining in the study by a given time as outcomes evolve over time. The inverse probability weighted data can be analyzed using methods appropriate for complete data, commonly generalized estimating equations (GEE; Mallinckrodt and Lipkovich, 2017). Therefore, IPW can easily be applied to categorical, time-to-event, or other non-normally distributed outcomes.

11.2.5 Considerations for Categorical and Time-to-Event Data

Many of the modeling considerations and principles regarding longitudinal analysis of continuous outcomes also apply to categorical outcomes. However, one key difference (discussed in Chapter 9) is regarding inference. Fixed-effect parameters estimated from a linear mixed-effects model in Gaussian data have both a marginal and hierarchical interpretation. In non-normal data, these are different targets of inference because the link between the mean of the outcome variable and the covariates is not identity. Therefore, separate models are needed for marginal and hierarchical inference in categorical data (Mallinckrodt and Lipkovich, 2017).

Incorporating an appropriate link function to account for nonlinearity and an appropriate distribution for the errors in categorical data increases computational complexity. This complexity limits the implementation of certain models, especially for likelihood-based estimation. Therefore, it is often useful to impute missing data using MI and then use GEE for analyses.

Most commonly used time-to-event analysis methods, such as the Kaplan–Meier approach and Cox regression, assume that the censoring times are independent of the event times (typically conditional on covariates observed at baseline) (Lipkovich, Ratitch, and O'Kelly, 2016). This so-called censoring at random assumption is referred to as ignorable or noninformative censoring. In censoring at random, for those at risk of an event at time t, the event hazard of those who are censored at time t is assumed to be equal to the hazard of those who are not censored. Methods that make this assumption ignore (do not model) the censoring mechanism because it would not provide additional information for inference about the survival process (Heitjan and Rubin, 1991).

Time-to-event analyses can fit into the same general analytic framework as previously discussed for continuous endpoints. For example, MI with time-to-event data is essentially the same as in the case of continuous or categorical data. Patients with censored times are considered as having missing values for event times, and the objective of the imputation process is to fill in these missing values.

11.3 Estimators for Treatment Policies That Were Not Included in the Trial or Not Followed

11.3.1 Introduction

Examples of estimands for this category include the difference between treatments regardless of adherence and without initiation of rescue medication. One specific example of this is what would have happened if patients took placebo after discontinuing the initially randomized study drug. Recently, a family of methods referred to as reference-based controlled imputation has emerged that are specific versions of pattern-mixture models designed to address these switches in treatment (see Mallinckrodt et al., 2019 and references therein). These methods trace back to an original publication by Little and Yau (1996).

In reference-based methods, the common idea is to construct a principled set of imputations that model specific changes in treatment. After relevant ICEs, values are multiply imputed by making qualitative reference to another arm in the trial. In other words, imputed post-ICE values have the statistical behavior of otherwise similar patients in a reference group. Initial implementations of reference-based imputation used MI (Carpenter, Roger, and Kenward, 2013). However, likelihood-based analogs have been proposed (Liu and Pang, 2016; Mehrotra, Liu, and Permutt, 2017).

Reference-based imputations can be tailored to specific scenarios. One variant of reference-based imputation, termed jump to reference (J2R), is implemented such that imputed values for patients in the experimental arm take on the attributes of the reference arm (placebo) immediately after ICEs. That is, the treatment benefit in patients in the active arm disappears immediately after a deviation from the intended treatment regimen (Carpenter, Roger, and Kenward, 2013; O'Kelly and Ratitch, 2014; Mallinckrodt and Lipkovich, 2017).

In a second approach, called copy reference (CR), the imputations result in a treatment effect that gradually diminishes after ICEs in accordance with the correlation structure implied by the imputation model. In a third approach, called copy increment from reference (CIR), the treatment effect observed when ICEs occurred is maintained by assuming the same slope or the same increments between visits as in the reference group (Carpenter, Roger, and Kenward, 2013; O'Kelly and Ratitch, 2014; Mallinckrodt and Lipkovich, 2017).

These methods, especially J2R, are more principled alternatives to addressing the same estimand as BOCF in continuous data and NRI in binary data. Using the placebo group as the definition of no benefit, reference-based imputation accounts for nonpharmacologic benefits, such as study effects and placebo effects, and thereby avoids the restrictive assumptions of zero change from baseline in placebo for BOCF and zero response rate for placebo in NRI (Mallinckrodt, Molenberghs, and Rathmann, 2017; Mallinckrodt and Lipkovich, 2017).

11.3.2 Reference-Based Approaches Using Multiple Imputation

To conceptually understand reference-based imputation, consider traditional MI based on MAR as involving a regression on residuals. Each patient's deviation from their respective group means prior to discontinuation (or more generally any relevant ICE) is used to impute the missing residuals. In MAR, imputed values are based on the imputed residual plus the mean of the group to which the patient was originally randomized.

In J2R, the regression on residuals is again applied. However, imputed values are based on adding the imputed residuals to the reference arm mean – not the arm to which the patient had been randomized (Carpenter, Roger, and Kenward, 2013; O'Kelly and Ratitch, 2014; Mallinckrodt and Lipkovich, 2017).

In CR, residuals are added to the reference arm mean as in J2R. However, in CR, the residuals are determined by the deviation of observed values from the reference arm – not the arm to which the patient was initially randomized, as is the case for MAR-based MI and J2R. This feature has the effect of giving credit to the drug arm for benefit resulting from the drug pre-ICE; that benefit declines over time in accordance with the correlations implied by the imputation model (Carpenter, Roger, and Kenward, 2013; O'Kelly and Ratitch, 2014; Mallinckrodt and Lipkovich, 2017).

In CIR, residuals are determined as in MAR-based MI and J2R. Missing values are imputed by adding the regressed residuals to hypothetical means such that the slope or differences between visits will be similar to those in the reference group (Carpenter, Roger, and Kenward, 2013; O'Kelly and Ratitch, 2014; Mallinckrodt and Lipkovich, 2017). Technical details for reference-based imputation based on MI, likelihood-, and Bayesian-based approaches are provided in Chapter 22.

It is also possible to construct "unconditional" imputations. In this approach, the imputed values are based on placebo group means and (possibly) individual patient baseline covariates. In other words, the post-baseline outcome history of patients is ignored.

The key assumption of reference-based imputation is that the reference arm describes the desired trajectory of the experimental arm after the occurrence of the ICEs, typically modeling a change in treatment. Sensitivity to departures from this assumption can be tested using delta-adjustment for MI-based approaches and mean delta-adjustment for the likelihood-based approaches (see Chapter 16 for a general discussion of sensitivity analyses and Chapter 23 for a detailed discussion of delta adjustment).

11.3.3 Considerations for Categorical and Time-to-Event Data

The ideas of reference-based imputation are extendable to categorical, count, and time-to-event data. It is useful to simply adopt the appropriate imputation model for the situation at hand, as in standard, MAR-based MI. Then, as in continuous data, the reference arm as the basis for imputing post-ICE values for experimental and control patients is used. For binary outcomes with imputation based on logistic regression models, patients in the experimental arm who discontinue early or have other relevant ICEs will have after discontinuation (or more generally after any relevant ICE) the logistic function for the reference group, which modifies the predicted probabilities for the classification levels. For an ordinal classification variable, patients in the experimental group who discontinue early or have other relevant ICEs have the cumulative logit function values of the reference group. For an imputed nominal classification variable, patients in the experimental group who discontinue early (or more generally have any relevant ICE) have the generalized logit model function values of the reference group (Carpenter and Kenward, 2013; van Buuren, 2007, 2018). For time-to-event endpoints, patients in the experimental group who are censored have the hazard function of the reference group (Lipkovich, Ratitch, and O'Kelly, 2016).

11.3.4 Likelihood and Bayesian Approaches to Reference-Based Imputation

Likelihood (Liu and Pang, 2016; Mehrotra, Liu, and Permutt, 2017) and Bayesian approaches (Liu and Pang, 2016) to reference-based imputation exist. Chapter 22 provides technical details on reference-based approaches using MI, likelihood, and Bayesian methods. In contrast to MI-based approaches that impute individual values to create multiple completed data sets for analysis, in likelihood and Bayesian approaches, a single analysis is used wherein means estimated from the data are used to implicitly impute data for those patients with irrelevant or missing data.

For example, assume the intent was to estimate mean change from baseline to endpoint assuming that patients who discontinue from the drug treatment group remained in the study taking placebo. The mean for the drug-treated group would be based on a mixture distribution of drug-treated patients who completed the trial and drug-treated patients who discontinued the randomized medication early whose distribution of outcomes is represented by otherwise similar placebo-treated patients.

This approach is conceptually like the J2R MI approach. As discussed in Chapter 22, replacing all missing or irrelevant values with a single mean in a single analysis has posed technical challenges in deriving appropriate estimates of uncertainty. In fact, subtle differences between likelihood- and MI-based approaches result in the two methods estimating slightly different estimands.

11.4 Summary

Although the number of potential hypothetical scenarios that might be considered is extensive, most of the practical applications fall into several categories: Outcomes that would have been expected if patients had not experienced relevant ICEs or what outcomes would have been if patients followed a specific treatment regimen, typically involving a change in treatments, that either was not available in the trial or was not followed by some patients if available.

Methods valid under the assumption of MAR, such as ML, MI, and IPW, are analytic options when assessing outcomes expected if ICEs had not occurred. Methods that assume MNAR, especially reference-based imputation, are analytic options for scenarios involving changes in treatment, such as when a treatment regimen was not available or not followed in a trial.

The key assumption for MAR-based analyses is that the missing data arise from an MAR mechanism. The key assumption of reference-based imputation is that the reference arm describes the desired trajectory of the experimental arm after the occurrence of the ICEs, typically modeling a change in treatment. Sensitivity to departures from these assumptions can be tested using delta adjustment for MI-based approaches and mean delta adjustment for the likelihood-based approaches.

12

Overview of Analyses for Principal Stratification Intercurrent Event Strategies

12.1 Introduction

The basic idea in principal stratification is to identify underlying strata (where strata are defined with respect to status for an intercurrent event (ICE) – ICE would/would not occur for the patient – under specific interventions) and then estimate causal effects within strata.

The ICH E9(R1) Addendum (2019), for example, proposes this strategy to account for the ICE of study discontinuation wherein inferences are drawn for the subset of patients that would adhere to all treatments in the study. Relevance of this stratum should be considered. For example, the subset of patients who would not dropout due to lack of efficacy on placebo may not be a suitable target for treatment. Other applications of principal stratification that may be useful include accounting for the ICE of death, wherein focus would be on the so-called survivor average causal effect, based on the stratum of patients that would survive on both (all) treatments. Another potentially meaningful application is vaccine trials where the treatment is intended to lessen the severity of symptoms in those who get infected regardless of treatment. However, at randomization, it is not known who will get infected. Focusing on the subset that actually did acquire infection may lead to bias because this is not a random subset and represents the outcome only under one intervention – vaccine or no vaccine. Therefore, principal stratification can be used to identify the stratum of patients who would have become infected on both (all) treatments.

In parallel group designs, only one potential outcome can be observed per patient. However, in many instances, it would be desirable to observe multiple potential outcomes; for example, the outcome of patient i on both the experimental and control treatments, such as would be possible in a cross-over design. Randomization facilitates inferences about the difference between experimental and control treatments when each patient receives only one treatment. However, post-randomization events (ICEs) can effectively break the randomization, thereby jeopardizing the validity of causal inferences about the randomized treatments. Principal stratification strategies can be useful for these situations, although additional identifying assumptions are needed for estimation.

12.2 Applications

In the context of accounting for ICEs, the principal stratum strategy targets the subset of the initially randomized population who would have the same status for an ICE (ICE would/would not occur) under a specific treatment regardless of the treatment arm to which they were actually randomized. For example, the population of interest might be the stratum of patients in which failure to adhere to treatments would not occur on either treatment, or the stratum of patients who would not need rescue medication on either treatment or the stratum who would remain alive on either treatment. Strata may also be defined with respect to ICEs under one treatment only. For example, a stratum of patients who would need rescue on placebo (regardless of whether they would need rescue on active). Estimands using this approach target a population of patients who need pharmacologic treatment.

Generalizability of strata should be considered (Ratitch et al., 2019a). For example, if focusing on the stratum compliant to both treatments, consider whether it is relevant to assess drug effects in a population whose illness is such that no patients on placebo would discontinue for lack of efficacy.

Other applications of principal stratification in clinical trials include situations in which it is not known at the time of randomization who will need the treatment. For example, in vaccine trials where the intervention is intended to reduce viral load, it is not known at randomization which patients will become infected. Principal stratification facilitates focus on the subset of all randomized patients who (are likely to) become infected regardless of vaccination. Another application of principal stratification is when dealing with death in evaluating quality of life. In this case, a meaningful treatment effect referred to as survivor average causal effect (SACE) is well defined in the principal stratum of patients who would have survived to a specific time point regardless of which treatment they were assigned to (Ratitch et al., 2019a).

The scientific question of interest frequently relates to the treatment effect only within the relevant stratum. However, one can also estimate the treatment effect within several (all) principal strata and then take a weighted average for an overall treatment effect. Membership in principal strata is not based on trial outcomes observed only on the actually assigned treatment, and this strategy therefore differs from subgroup or completers analyses. If members of the principal stratum are correctly identified and assessed for the entire trial duration, statistical analyses for this estimand are straightforward (ICH, 2019).

Identifying strata membership in advance of randomization can be difficult because the occurrence of ICEs is hard to predict using baseline covariate alone (ICH, 2019). Examples of designs that can help target appropriate patients include enrichment designs or run-in and randomized withdrawal designs (ICH, 2019). In a parallel group design, membership in a principal stratum must usually be inferred from covariates (ICH, 2019). If the targeted population is not correctly identified, restricting analyses to patients with a specific status with respect to the ICEs confounds treatment with patient characteristics that influence the probability of ICEs, thereby systematically excluding different patients who experience differential rates and types of ICEs on different treatments. Consequently, it is necessary to assess robustness of conclusions to the assumptions, in this case stratum membership, using appropriate sensitivity analyses (ICH, 2019).

The basic idea in principal stratification is to identify underlying strata (where strata correspond to status for an ICE – ICE would/would not occur for the patient) and then compute causal effects only within strata. It is a generalization of the local average treatment effect (LATE) (Imbens and Angrist, 1994; ICH, 2019). Assume the ICE of interest is the binary outcome of adherence to study medication (yes or no) and that treatment is also binary, with an experimental group and control. This example has four strata:

1. Those who would adhere to both treatments
2. Those who would not adhere to either treatment
3. Those who adhere to the experimental treatment but not the control treatment
4. Those who adhere to the control treatment but not the experimental treatment

If the stratum for each patient were known, then outcomes could be compared within the stratum of interest and a valid estimate of the causal treatment effect could be obtained from basic statistical analyses (ICH, 2019). The fundamental analytic problem is that stratum membership is not known, resulting in the need for specialized analyses, modeling assumptions, and sensitivity analyses (see Chapter 24 for technical details on principal stratification analyses).

12.3 Summary

The basic idea in principal stratification is to identify underlying strata (where strata are defined with respect to status for an ICE – ICE would/ would not occur for the patient – under specific interventions) and then estimate causal effects within strata. In parallel group designs, only one potential outcome can be observed per patient although it would often be desirable to observe multiple potential outcomes. Randomization facilitates inferences about the difference between experimental and control treatments when each patient receives only one treatment. However, post-randomization events (ICEs) can effectively break the randomization, thereby jeopardizing the validity of causal inferences about the randomized treatments. Principal stratification strategies can be useful for these situations although additional identifying assumptions are needed for estimation.

13

Overview of Analyses for While-on-Treatment Intercurrent Event Strategies

The authors credit James Bell as co-author through his contributions to the paper upon which this Chapter is based.

In the while-on-treatment strategy of accounting for intercurrent events (ICEs), response to treatment prior to the occurrence of relevant ICEs is the primary focus. For repeated measures variables, values after an ICE are considered not meaningful. For example, patients with a terminal illness may discontinue a symptomatic treatment because they die, yet the success of the treatment before death is still relevant (ICH, 2019).

An easily interpreted while-on-treatment strategy is a slope analysis based on measurements prior to any ICE via a linear mixed model (Mallinckrodt and Lipkovich, 2017). If trajectories over time are approximately linear, then a linear model with random effects for the slope and intercept may be used to model these longitudinal measurements. An alternative is a two-stage approach where in step one a slope is calculated for each patient and in step two the slopes are fit as the dependent variable in an analysis of variance (ANOVA) or analysis of covariance (ANCOVA). A comparison of the slopes between treatment arms in the mixed-effects model simultaneously addresses both hypothetical and while-on-treatment estimands. The two-step method is purely a while-on-treatment approach.

The time independence of linear slopes makes this an attractive method of addressing while-on-treatment estimands where treatment is expected to result in a disease-modifying effect, so long as the strong assumption of linearity is reasonable. Typically for symptomatic treatments, no or little difference in slope can be expected (although there may be differences in the intercept). Although not common in practice, this model could be extended to include higher order terms.

Other while-on-treatment analyses include rates per unit time and change to last observation (ICH, 2019). Rates per unit time are commonly used to assess adverse events. For example, the rates of serious infections may be expressed as x.x% per 100 patient years exposure. Most common analysis models for recurrent-event data, including the negative binomial model, assume a constant hazard over time and hence lend themselves to while-on-treatment approaches.

Assessment-of-treatment effects in the last observation approach are based on comparing means or percentages using the last observation, regardless of when that occurred. Every patient has a last observation. Therefore, the while-on-treatment approach yields no missing data due to relevant ICEs. An analysis of variance (covariance) model with treatment as a factor is an appropriate statistical analysis. Last observation analyses may be useful when it is reasonable to assume patient outcomes were stable at last observation.

Treatment effects can be based on averaging values prior to the occurrence of relevant ICEs. Intermittent missing measurements need to be interpolated or imputed based on plausible assumptions that account for the uncertainty due to missing data (ICH, 2019). Multiple imputation based on the missing at random assumption is a plausible approach for intermittent missing data. After imputation, the completed data sets can be analyzed via ANOVA or ANCOVA.

The interpretation of last observation carried forward (LOCF)-based approaches should be considered because there is a danger of unfair comparisons. For instance, where treatments aim to maintain status or reduce progression over time, the arm with more ICEs will appear to have relatively higher efficacy because the last observations for patients with an event will on average be better than if observations continued until the final visit.

14

Overview of Analyses for Treatment Policy Intercurrent Event Strategies

The authors credit James Bell as co-author of this chapter through his contributions to the paper upon which this Chapter is based.

In the treatment policy strategy, the observed values for the variable of interest are used regardless of whether an intercurrent event (ICE) occurred; that is, no ICE is a departure from the treatment regimen of interest. Adopting a treatment policy strategy across all ICEs has become known as a "pure ITT approach," and its primary usage has been motivated by its preservation of randomization. The treatment policy strategy can only be applied if it is possible, at least in principle, to observe the outcome as planned. This strategy is not applicable, for example, when the ICE is death. Since death is always at least a possibility in a trial, the treatment policy strategy is technically undefinable unless death is the primary outcome (ICH, 2019). However, this strategy may still be relevant when few deaths are expected, with death treated as missing (using, essentially, a hypothetical strategy) (Mallinckrodt et al., 2019).

The pure treatment policy strategy broadens the treatment regimens under evaluation because whatever treatment is taken or not taken, for however, long it is taken, is part of the assessed treatment regimen. As discussed in detail in Chapter 3, loosely defined comparisons may not be meaningful because the target of inference is not specifically defined, and the treatment regimens used in blinded and placebo-controlled clinical trials may not correspond to real-world practice (Ratitch et al., 2019a). The treatment policy is perhaps more productively used for specific types of ICEs in conjunction with specific treatment regimens, with other strategies used to account for other ICEs (Ratitch et al., 2019a).

However, if the treatment policy strategy is the only approach used to account for all ICEs, that is, all ICEs are ignored, then the remaining analytic consideration is that some data will inevitably be missing due to some patients missing study assessments intermittently or discontinuing from the study prematurely. When handling missing data under the treatment policy strategy, it is important that the methods used condition upon the occurrence of ICEs to reflect the value that would have been observed. For instance, an imputed value for a patient with missing data following rescue should correctly reflect that the patient took the rescue therapy. The imputation model for that patient may therefore differ from the imputation model applied to patients with missing data who did not take rescue. Since ICEs reflect changes of treatment, it must be assumed that treatment effects may change following an ICE such that simple time-dependent covariate adjustments are insufficient. Multiple imputation-based techniques such as reference-based imputation can be used for these analyses (see Chapter 22 for more details) using either the placebo arm data or observed post-ICE data as the source of imputation information.

15

Missing Data

15.1 Introduction

Missing data is not an intercurrent event (ICE); it is a consequence of ICEs. For example, early-study termination may follow an ICE of premature study treatment discontinuation and because of that ICE outcomes are missing. Chapter 4 noted that the composite and while-on-treatment strategies to handling ICEs can yield no missing data. However, in the treatment policy, in certain hypothetical strategies, and in some principal strata, missing data is an important analytic consideration. This chapter provides an overview of key concepts relating to missing data and how to account for it in analyses.

15.2 Basic Principles

The best way to deal with missing data is to prevent it (NRC, 2010; Fleming, 2011; Mallinckrodt, 2013; O'Kelly and Ratitch, 2014). However, despite our best efforts to minimize missing data, incompleteness is inevitable. The fundamental problem caused by missing data for relevant estimands is that the balance provided by randomization is lost if, as is usually the case, the patients with missing data have different outcomes from those who complete the study. This imbalance can lead to biases in the comparison of the treatment groups, along with loss of power for hypothesis testing and reduced precision in parameter estimates.

Missing data in longitudinal clinical trials is a complex and wide-ranging topic, in part because missing data may arise in many ways. Intermittent missing data occurs when patients miss scheduled assessments but attend subsequent visits. Dropout (withdrawal, attrition) is when patients miss all subsequent assessments after a certain visit. Intermittent missing data is generally less problematic than withdrawal/attrition because having observed data bracketing the unobserved value makes it easier to verify assumptions about the missing data than when no subsequent values are observed.

Incomplete data must be addressed in the analyses. However, addressing missing data requires unverifiable assumptions because we simply do not have the missing data about which the assumptions are made (Verbeke and Molenberghs, 2000). Despite the occurrence of incomplete data, evaluation of relevant estimands may be valid provided the statistical methods used are sensible. Carpenter and Kenward (2007) define a sensible analysis as one where:

1. The variation between the intervention effect estimated from the trial data and that in the population is random. In other words, trial results are not systematically biased.

2. As the sample size increases, the variation between the intervention effect estimated from the trial data and that in the population gets smaller and smaller. In other words, as the size of the trial increases, the estimated intervention effect more closely reflects the true value in the population. Such estimates are called consistent in statistical terminology.

3. The estimate of the variability between the trial intervention effect and the true effect in the population (i.e., the standard error) correctly reflects the uncertainty in the data.

Valid inferences for relevant estimands are possible if these conditions hold despite the missing data. However, the analyses required to meet these conditions may be different from the analyses that satisfy these conditions when no values are missing. Regardless, whenever outcomes intended to be collected are missing, information is lost, and estimates are less precise than if data were complete (Mallinckrodt, 2013).

When drawing inferences from incomplete data, it is important to recognize that the potential bias from missing data can either mask or exaggerate the true difference between treatments for the relevant estimand (Mallinckrodt et al., 2008; NRC, 2010). Moreover, the direction of bias has different implications in different scenarios. For example, underestimating treatment differences in efficacy is bias *against* an experimental treatment that is superior to control but is bias *in favor* of an experimental treatment that is inferior to control. This situation has particularly important inferential implications in non-inferiority testing. Underestimating treatment differences in safety is bias in favor of an experimental treatment that is less safe than control but is bias against the experimental drug that is safer than control (Mallinckrodt and Lipkovich, 2017).

15.3 Missing Data Mechanisms

15.3.1 General Considerations

To understand the potential impact of missing data on a relevant estimand and to choose an appropriate analytic approach for a situation, the stochastic process(es) (i.e., mechanisms) leading to the missingness must be considered. The following taxonomy of missing data mechanisms is now well established in the statistical literature (Little and Rubin, 2002).

Data is missing completely at random (MCAR) if, conditional upon the covariates (e.g., treatment group, baseline severity, investigative site) in the analysis, the probability of missingness does not depend on either the observed or unobserved outcomes of the variable being analyzed.

Data is missing at random (MAR) if, conditional upon the covariates in the analysis and the observed outcomes of the variable being analyzed, the probability of missingness does not depend on the unobserved outcomes of the variable being analyzed.

Data is missing not at random (MNAR) if, conditional upon the covariates in the analysis model and the observed outcomes of the variable being analyzed, the probability of missingness does depend on the unobserved outcomes of the variable being analyzed.

Another way to think about the distinction between MAR and MNAR is that if, conditional on observed outcomes, the statistical behavior (means, variances, etc.) of the unobserved data is similar to the behavior of the observed data, then the missingness is MAR, if not, then MNAR. In MCAR, the outcome variable is not related to the probability of dropout (after conditioning on baseline covariates). In MAR, the observed values of the outcome variable are related to the probability of dropout, but the unobserved outcomes are not related to the probability of dropout after conditioning on covariates and observed outcomes. In MNAR, the unobserved outcomes are related to the probability of dropout even after conditioning on the observed outcomes and covariates.

Mallinckrodt et al. (2008) summarized several key points that arise from the precise definitions of the missingness mechanisms given above. First, the definitions are all conditional on the model. Therefore, characterization of the missingness mechanism does not rest on the data alone; it involves both the data and the model used to analyze them. Consequently, missingness that might be MNAR given one model could be MAR given another. Moreover, as the relationship between the dependent variable and missingness is a key factor in the missingness mechanism, the mechanism may vary from one outcome to another within the same data set.

Moreover, when dropout rates differ by treatment group, it would be incorrect to conclude on these grounds alone that the missingness mechanism was MNAR and that analyses assuming MCAR or MAR were invalid. If dropout depended only on treatment, and treatment was included in the model, the mechanism giving rise to the dropout was MCAR. This is one example of what some have termed covariate-dependent MCAR (Little, 1995; O'Kelly and Ratitch, 2014). The distinction between covariate-dependent MCAR and MCAR applies to all baseline covariates, not just treatment.

Given that the missingness mechanism can vary from one outcome to another in the same study and may depend on the model and method, statements about the missingness mechanism without reference to the model and the variable being analyzed are problematic to interpret. This situational dependence also means that broad statements regarding missingness and validity of analytic methods across specific disease states are unwarranted (Mallinckrodt, 2013). In addition to MCAR/MAR, ignorability rests on the separation of the parameters describing the outcome process and parameters describing the missingness process (Verbeke and Molenberghs, 2000) (see Chapter 18 for more details).

15.3.2 Considerations for Time-to-Event Analyses

Terms such as ignorable missingness, which are often associated with time-to-event analyses, can also be problematic to interpret. For example, with likelihood-based estimation, if certain conditions hold, the missingness is ignorable if it arises from an MCAR or MAR mechanism but is nonignorable if it arises from an MNAR process (Verbeke and Molenberghs, 2000). In this context, ignorable means the modeling of the missing data mechanism is unnecessary because the observed data combined with a model for the outcome yield unbiased estimates of the parameters governing the measurement process. However, if other forms of estimation are used, missing data may be ignorable only if arising from an MCAR mechanism. Hence, describing missing data as ignorable or nonignorable requires reference to both the estimation method and the analytic model besides the missing data mechanism (Mallinckrodt, 2013). As with continuous outcomes, besides MCAR/MAR, ignorability rests on the separation of the parameters describing the outcome process and parameters describing the missingness process.

Informative censoring is yet another term used to describe the attributes of missing data, typically used in the context of time-to-event analyses. If the response variable was time to an event, patients not followed long enough for the event to occur have their event times censored at the time of last assessment. Time-to-event analyses often assume that causes of censoring are independent of what causes events, typically after conditioning on baseline covariates or strata, such as in a Cox proportional hazard regression model. A term similar to that in the context of longitudinal data would be censoring at random. When the likelihoods of outcomes and censoring are related, conditional on covariates, then censoring not at random exists. When the likelihoods of outcomes and censoring are not related, conditional on covariates, then censoring is at random. For example, if patients discontinue because of poor response to treatment but did not yet have the event of interest, the censoring indirectly reflects bad outcomes and is not at random.

Important considerations differentiate censoring at random in the time-to-event context from missing at random in the repeated measures setting. In the time-to-event setting, typically only one observation per patient is available (except in recurrent event settings); no intermediate observations exist. In longitudinal settings with repeated measures, most patients have at least partially observed sequences upon which to condition the results. For censoring to be not at random, it must be in addition to the information from the partially observed outcomes and covariates. With no intermediate observations in the time-to-event setting, it is more likely censoring would be not at random than in an otherwise similar scenario where analyses can condition on the intermediate observations. However, in the repeated measures setting, the partially observed outcomes, although predictive of unobserved outcomes, do not exclude the possibility that the unobserved outcome could take on any value. In contrast, censoring in a time-to-event setting does exclude the possibility of outcomes. For example, if a patient is censored at month 6, it is known that the unobserved event time is not less than 6 months.

15.4 Analytic Considerations

15.4.1 General Considerations

Until recently, guidelines for the analysis of clinical trial data provided only limited advice on how to handle missing data, and analytic approaches tended to be simple and ad hoc (Molenberghs and Kenward, 2007). Simple and ad hoc methods became popular during a time of limited computing power because they restored the intended balance to the data, allowing implementation of the simple analyses for complete data. However, with the seminal work on analysis of incomplete data by Rubin (1978a), including the now common taxonomy of missingness mechanisms, attention began to shift to accounting for the potential bias from missing data.

The analytic conundrum missing data poses is that MCAR is usually not a reasonable assumption; MAR may be reasonable, but there is no way to know for certain that it holds, and there is no way to be certain that an MNAR method and model is appropriate (Mallinckrodt, 2013). A sensible and useful compromise between blindly shifting to MNAR models and ignoring them altogether is to use MNAR methods in sensitivity analyses (Molenberghs and Kenward, 2007; Mallinckrodt et al., 2008). Therefore, for relevant estimands, primary analyses of longitudinal clinical trials should often be based either on methods that assume MAR or that make specific clinically plausible MNAR assumptions (e.g., for estimands involving a hypothetical strategy), and robustness of the primary result should be assessed using sensitivity analyses (Verbeke and Molenberghs, 2000; Molenberghs and Kenward, 2007; Mallinckrodt et al., 2008; Siddiqui, Hong, and O'Neill, 2009; NRC, 2010; Mallinckrodt et al., 2014).

Common analytic frameworks under the MAR assumption include direct likelihood analyses, multiple imputation, inverse probability weighting, and augmented inverse probability weighted analyses (Mallinckrodt and Lipkovich, 2017). These analyses were introduced in Chapter 11, and technical details are provided on each analytic approach in Part IV.

15.4.2 Considerations When Changing Treatment Is Possible

In the MAR framework, previous outcomes and covariates are used to predict what the unobserved outcomes would have been if they had been observed. The explicit assumption is that the observed data can predict the unobserved data. When the treatment regimen of interest is such that adding a treatment or switching to another treatment is an ICE that marks a deviation from the regimen of interest, then data after switching or adding a treatment is not relevant and should not be included in the analysis. Validity of MAR analyses in these situations hinges, as is more generally the case, on whether covariates and previous outcomes explain the reason for adding or switching treatment.

When the treatment regimen of interest does not allow patients to switch or add treatments, the switching must be explicitly addressed in the analysis. To illustrate, consider the six patient profiles in Table 15.1. If the target of inference is the initially randomized treatment, and the estimand of interest is the difference between treatment means at Visit 3 if patients had adhered to treatment, the only data that are relevant are visits denoted with an X. An

TABLE 15.1

Example Patient Profiles

1	X X X	Randomized treatment alone through the end of study
2	X ‡ ‡	Randomized treatment at V1; treatment plus rescue at V2, V3
3	X + +	Randomized treatment at V1; rescue only at V2, V3
4	X O O	Randomized treatment at V1; rescue alone at V2, V3
5	X ‡ O	Randomized treatment at V1; randomized treatment + rescue at V2; rescue at V3
6	X + O	Randomized treatment at V1; randomized treatment + rescue at V2; no treatment at V3

Note: X = randomized treatment; O = no treatment (with the randomized treatment, experimental or placebo, discontinued, and no other treatment started); ‡ = randomized treatment with an addition of concomitant rescue; and + = rescue treatment without the randomized treatment.

estimate for this estimand can be obtained using an analytic method under MAR, which uses the relevant data to estimate what would happen under complete adherence to the initial treatments; that is, what would happen if all visits for all patients had been observed while adhering to the initially randomized treatment. However, if the target of inference is the treatment regimen involving the initially randomized treatment and rescue, then although there is less missing data, the situation is more complex.

For example, consider profile 4 where patients are observed on randomized treatment only at V1, with V2 and V3 missing. Should the goal of the analysis be to estimate what would have happened if these patients had taken randomized treatment at V2 and V3 (Xs at V2 and V3), or should the goal be to estimate what would have happened if the patients had taken randomized treatment plus rescue at V2 and V3 (‡ at V2 and V3), or should the goal be to estimate what would have happened if these patients had taken rescue only at V2 and V3 (+ at V2 and V3)?

For profile 5, observations are taken under randomized treatment at V1 and randomized treatment plus rescue at V2, with V3 missing, should the goal of the analysis be to estimate what would have happened if these patients had taken randomized treatment plus rescue at V3 (‡ at V3), or should the goal be to estimate what would have happened if these patients had taken rescue only at V3 (+ at V3)?

Importantly, the pure ITT approach wherein the treatment policy strategy is used for all ICEs does not provide clear guidance on how to handle the various missing data possibilities. This is an illustration of the difficulties in the treatment policy strategy with loosely or flexibly defined treatment regimens. Specifically defined treatment regimens, including the sequence of treatments and reasons for changing treatments, are needed to guide the proper handling of the missing data.

As will be discussed in Chapter 19, multiple imputation provides a flexible framework for handling the various profiles, whereas direct likelihood and IPW may lack the necessary flexibility depending on exactly how the missing data is to be handled.

15.5 Inclusive and Restrictive Modeling Approaches

Collins, Schafer, and Kam (2001) describe restrictive and inclusive modeling philosophies. Restrictive models typically include only the design factors of the experiment and perhaps one or a few covariates. Inclusive models include, in addition to the design factors, auxiliary variables whose purpose is to improve the performance of the missing data procedure.

Recalling the specific definition of MAR provides the rationale for inclusive modeling. Data is MAR if, conditional upon the variables in the model, missingness does not depend on the unobserved outcomes of the variable being analyzed. Therefore, if additional variables are added to the model that explains missingness, MAR may be valid, whereas if the additional variables are not included the missingness would be MNAR.

Ancillary variables can be included in likelihood-based analyses by either adding the ancillary variables as covariates or as additional response variables to create a multivariate analysis. However, the complexity of multivariate analyses and the features of most commercial software make it easier to use ancillary variables via MI. With separate steps for imputation and analysis, post-baseline, time-varying covariates – possibly influenced by treatment – can be included in the imputation step of MI to account for missingness but then not included in the analysis step to avoid confounding with the treatment effects, as might be the case in a likelihood-based analysis.

15.6 Summary

For some estimands, missing data is an incessant and complex problem both in longitudinal and (recurrent) event clinical trials. The best way to deal with missing data is to prevent it. However, some degree of incompleteness is inevitable. The potential impact of missing data depends on the underlying mechanism. In MCAR, the probability of dropout is not related to either the observed or the unobserved outcomes. This situation is unlikely in most clinical trial scenarios. Two ways to think about the less restrictive MAR mechanism is (1) the missingness does not depend on missing data given observed data and (2) the distribution of unobserved future outcomes is the same as the distribution of observed future outcomes, conditional on earlier outcomes. If these conditions do not hold, then data is MNAR. Analytic methods based on MAR include maximum likelihood, multiple imputation, inverse probability weighting, and augmented inverse probability weighting. MAR-based analyses can employ inclusive or restrictive modeling philosophies. Both with MAR and MNAR primary analyses, it is important to assess the degree to which inferences may be influenced by departures from the untestable assumptions.

16

Sensitivity Analyses

16.1 General Considerations

Although most of the initial attention regarding the ICH E9(R1) Addendum (ICH, 2019) has been on estimands, the document also covers sensitivity analyses. The addendum states that the purpose of sensitivity analyses is to assess whether inferences based on an estimand are robust to limitations in the data and deviations from the assumptions used in the statistical model for the main estimator. The addendum further notes the statistical assumptions underpinning the main estimator should be documented, and that one or more analyses focused on the same estimand should be prespecified as sensitivity analyses. The addendum recommends that sensitivity analysis should not alter many aspects of the main analysis simultaneously. It is preferable to alter one aspect at a time in order to clearly identify how each assumption impacts robustness (ICH, 2019).

The addendum also draws distinction between testable and untestable assumptions. The focus of this chapter is mostly on untestable assumptions because this is where sensitivity analyses are needed. Assumptions such as normality or linearity are testable. When these assumptions are proven to be invalid, an alternative statistical method or model will clearly be needed and usually readily found. For example, if normality assumptions are violated, a nonparametric method can be used instead, and inferences can be based on the alternative analysis. Alternatively, data transformations can be considered. In contrast, with untestable assumptions, there is no objective basis upon which to select or compare models; therefore, the goal of sensitivity analyses is to quantify to what degree inferences from the primary analysis are influenced by *plausible* departures from assumptions (ICH, 2019).

Emphasis is placed on plausibility because considering implausible scenarios could be more misleading than informative. For example, consider a clinical trial investigating a drug hoped to alleviate pain where a visual analog scale from 0 to 100 is the primary outcome. Zero indicates no pain, and 100 indicates the worst possible pain. Assume pain ratings at baseline range from 40 to 70 and that the worst rating from any patient at any time point is 75. Depending on what estimand is being evaluated, it might be reasonable to assume in sensitivity analyses that missing data take on unfavorable values, but it would not be useful to assume the missing values took on the worst possible value of 100 because no patients had observed values close to that outcome value.

The addendum notes that missing data requires attention in sensitivity analysis because the assumptions underlying any method may be hard to justify and may not be testable (ICH, 2019). However, given that the addendum focuses heavily on extending previous discussions on missing data into the broader context of handling of intercurrent events (ICEs), so too should sensitivity analyses move beyond assessing the impact of missing data assumptions to more broadly assessing sensitivity to methods of accounting for ICEs.

Given the newness of ideas around and methods for handling ICEs, ideas and methods for using sensitivity and supplementary analyses to assess assumptions made in the various strategies for handling ICEs are currently lacking. Some initial ideas are outlined in this chapter for assessing sensitivity in each of the strategies for handling ICEs. However, more thought, research, and development of new methods are needed.

The following sections in this chapter first distinguish supplementary analyses from sensitivity analyses and then address topics in assessing sensitivity to missing data assumptions, followed by initial ideas on assessing sensitivity for each of the five strategies for handling ICEs.

16.2 Supplementary Analyses

The ICH E9(R1) Addendum distinguishes sensitivity analyses from supplementary analyses. It states that interpretation of trial results should focus on the primary estimand and primary estimator, whereas supplementary analyses targeting different estimands play a secondary role in order to more fully understand the treatment effects (ICH, 2019). Supplementary analyses can include different estimands that involve the primary endpoint and/or the primary estimand based on secondary endpoints.

An example of testing different estimands for the primary clinical outcome measure includes supplementing a mean change analysis of a continuous outcome with an analysis that compares the percentage of patients with clinically relevant improvements on that same measure (responder rate analysis) (ICH, 2019). Similarly, supplementary analyses could assess estimands for a treatment regimen involving the initially randomized medication in addition to the primary estimand that focuses only on the initial medications. Examples of secondary endpoints for the same estimand include supplementing a symptom severity primary estimand with quality of life and functional endpoints or secondary symptoms such as changes in sleep disturbances in patients being treated for pain.

An unfocused approach to supplementary analyses can yield confusion from trying to draw inference from a multitude of results that arise from testing many combinations of estimands and endpoints. Therefore, a focused approach to supplementary analyses should be driven through an under-standing of what important scientific questions or what aspects of sensitivity remained unaddressed from the primary estimand, estimator, and sensi-tivity analyses in order to construct a parsimonious set of supplementary analyses. Supplementary analyses may be especially useful for alternative estimands such as when addressing the differing preferences in estimands across diverse stakeholders.

Supplementary analyses can also be used to understand the impact of data limitations, such as protocol violations. The ICH Addendum (ICH, 2019) indicates that historically analysts have included results based on both the full analysis set and a subset of patients who had no meaningful protocol violations – the so-called per-protocol dataset (PPS). Although consistent results from analyses based on the full analysis set and per-protocol set may increase confidence in the trial results, the ICH Addendum (ICH, 2019) notes that PPS might be subject to severe bias due to loss of randomization. The addendum goes on to note that some protocol violations may be addressed as ICEs. Where most proto-col violations are handled through the construction of the estimands, the number of remaining violations will be low, and analysis of the PPSs might not add additional insights (ICH, 2019). Therefore, analysts should consider whether per-protocol analyses add meaningful information rather than simply following custom and including a set of per-protocol analyses.

16.3 Assessing Sensitivity to Missing Data Assumptions

16.3.1 Introduction

It is important to keep missing data sensitivity analyses within the overall context of assessing uncertainty in clinical trial results. This uncertainty arises from different sources: (1) inherent imprecision in parameters of the model estimated from a finite sample, (2) model selection (e.g., for the mean and covariance functions), and (3) the level of uncertainty due to incomplete data. Sources 1 and 2 can be assessed from the observed data. What sets missing data apart is that not all the uncertainty from incompleteness (missing data) can be objectively evaluated from the observed data and hence the need for missing data sensitivity analyses (Molenberghs and Kenward, 2007).

As noted in Chapter 15, missing data is typically an important consider-ation for treatment policy and hypothetical strategies to dealing with ICEs. When data are missing, the conundrum faced by analysts is that a missing completely at random (MCAR) mechanism is almost certainly not valid, and the assumption of missing at random (MAR) and the specific assumptions made by missing not at random (MNAR) models are not testable (Molenberghs and Kenward, 2007; NRC, 2010). Therefore, it is not possible to distinguish MAR from MNAR in practice (Verbeke and Molenberghs, 2000; NRC, 2010). A data-based distinction between MAR and MNAR is not possible because the fit of any MNAR model can be reproduced exactly by an MAR counterpart (Molenberghs and Kenward, 2007). Of course, such a pair of models will produce different predictions of the unobserved outcomes and therefore different estimates of the treatment effect, given the observed outcomes.

Therefore, no statistical *solution* to the problems caused by missing data is possible. However, assuming MAR for the primary estimator, especially when combined with trial design and conduct to minimize missing data, is often a useful starting point, with sensitivity analyses focusing on robustness to departures from MAR (Verbeke and Molenberghs, 2000; Molenberghs and Kenward, 2007). The remainder of this section focuses on assessing sensitivity to departures from MAR.

16.3.2 Assessing Sensitivity to Departures from MAR

Sensitivity analyses can be a single analysis or a series of analyses with differing assumptions. The aim is to compare results across sensitivity analyses to make clear how much inferences rely on the assumptions (NRC, 2010; Phillips et al., 2016). In clinical trials, sensitivity analyses typically focus on inferences regarding the treatment effects. Therefore, the primary aim of sensitivity analyses is to assess how treatment effects vary depending on assumptions about the missing data (Mallinckrodt and Lipkovich, 2017; Mallinckrodt et al., 2014). More specifically, the aim of sensitivity analyses for missing data is typically to evaluate the degree to which inferences for the relevant estimand are influenced by departures from MAR.

Unlike parameters estimated from observed data, there is no information in the observed data about the missing data sensitivity parameters. Therefore, analysts typically choose a plausible range of scenarios or values for sensitivity parameters to create relevant departures from MAR. The MNAR results are then compared to the MAR result (Carpenter, Roger, and Kenward, 2013).

Two basic routes can be considered in assessing the robustness of inferences to departures from MAR. Results from multiple MNAR models can be compared. However, inferences in this context are difficult. Results may differ between MNAR models because both models are wrong, or because one model is wrong, but we cannot know which one, or results can vary simply due to chance differences (Permutt, 2015b). The preferred approach is to add a sensitivity component or parameter(s) to the otherwise similar primary analysis and vary the sensitivity (MNAR) parameter(s) across a plausible range (Mallinckrodt and Lipkovich, 2017; ICH, 2019).

The controlled-imputation family of methods, which includes reference-based imputation and delta adjustment, has gained favor as sensitivity analyses because their assumptions are transparent and easy to understand (O'Kelly and Ratitch, 2014; Mallinckrodt and Lipkovich, 2017). In other words, it is easy to define and understand how the sensitivity analysis deviates from MAR. In delta adjustment, a sensitivity parameter termed delta is varied across a range of values that progressively departs further from MAR. This has been termed a tipping-point approach. If the treatment effect remains significant across plausible departures from MAR, inferences are not sensitive to plausible departures from MAR. This approach can be applied to continuous outcomes where treatment effects are assessed as mean changes, to time-to-event outcomes where treatment effects are assessed as hazard rates or ratios, and for categorical outcomes where treatment effects are assessed as percentages. A similar approach can be considered to ascertain values of delta for which the treatment effect does or does not retain a specific degree of clinical relevance (ICH, 2019) (see Chapter 23 for technical details and example implementations of delta-adjustment analyses).

Alternatively, a worst plausible departure from MAR can be defined using hypothetical strategies for dealing with the ICEs that give rise to missing data. For example, reference-based imputation can be used to assume that patients who discontinued their treatment or treatment regimen had no benefit from the treatment. This is accomplished by imputing values as if the patient were in the placebo group after breaking from the assigned treatment. As with delta adjustment, reference-based imputation can be implemented for continuous, time-to-event, and categorical outcomes.

Use of delta adjustment is not limited to assessing departures from MAR. Delta adjustment can also be used in combination with reference-based imputation approaches to assess departures from specific MNAR conditions (see Chapter 22 for technical details and example implementations of reference-based imputation).

16.4 Sensitivity to Methods of Accounting for Intercurrent Events

16.4.1 Introduction

At the time of this writing, we are not aware of literature that explicitly addresses assessing sensitivity to the means of handling ICEs. Therefore, the ideas presented in this section should be taken as initial suggestions and as inspiration for further research. The ICH E9(R1) (ICH, 2019) notes that sensitivity analyses should assess both limitations in the data and modeling assumptions. For methods of handling ICEs that rely on statistical models, particularly the handling of missing data (hypothetical, principal stratification, treatment policy), assessing sensitivity to modeling assumptions is important. For methods to handle ICEs that yield no missing data (composite, while-on-treatment), data limitations and how outcomes are defined when ICEs occur, deserve attention.

16.4.2 Sensitivity Analyses for Hypothetical Strategies

Hypothetical strategies yield an estimate for the particular scenario under investigation only. The scenario must be well specified, including what patients are assumed to do and assumed not to do (ICH, 2019). Therefore, sensitivity analyses are important in assessing robustness to departure from the assumed conditions. The previous section focused on missing data sensitivity analyses. One of the examples discussed was for assessing a hypothetical estimand of the treatment effects if all patients adhered to their initial treatment or treatment regimen. However, many ideas from that section could be applied to other hypothetical estimands.

For example, consider a *de facto* (effectiveness/as actually taken) primary estimand that is evaluated assuming patients who discontinue study medication have subsequent values that mimic otherwise similar placebo-treated patients. This assumption is essentially assigning no benefit for patients who do not adhere to the experimental drug. Sensitivity to the imputation, based on the assumption that otherwise similar placebo patients describe no benefit for drug-treated patients who discontinue, can be assessed using the delta-adjustment tipping point approach as described in the previous section for assessing sensitivity to departures from MAR. One of the advantages of delta adjustment is that it can be applied to any hypothetical setting.

16.4.3 Sensitivity Analyses for Composite Strategies

As noted in Chapter 4, composite strategies may in various manners incorporate adherence with efficacy or functional outcomes. This entails an implicit assumption that adherence decisions in the trial are like those in clinical practice (Mallinckrodt, Molenberghs, and Rathmann, 2017; Mallinckrodt and Lipkovich, 2017). This consideration is especially important in trials with placebo, randomization, and/or blinding because these factors are never present in clinical practice (Mallinckrodt, Molenberghs, and Rathmann, 2017; Mallinckrodt and Lipkovich, 2017).

We do not see a straight-forward statistical approach to directly assessing sensitivity to this clinical assumption. However, analysts should consider assessing sensitivity to how outcomes are defined for patients with ICEs. For example, rather than assigning unfavorable outcomes for all patients who discontinue study drug, failure can be assigned to only those patients who discontinued for reasons related to the study drug, such as lack of efficacy or adverse events. Outcomes for patients who discontinue for reasons not related to study drug could be based on data observed until the point of discontinuation and/or missing outcomes predicted based on the observed data.

Moreover, for composite strategies, it is also useful to assess sensitivity to the actual outcomes that are assigned. For example, if patients who discontinue study drug are arbitrarily assigned an unfavorable value for a continuous value, that value should be specified and justified in the protocol with prespecified sensitivity analyses that include alternative values. Similar considerations apply to ranked outcomes. For example, one might assume that early discontinuation is a worse outcome than any observed outcome, but this is an assumption and sensitivity to it should be investigated. Furthermore, if it is assumed that all patients with a relevant ICE are treatment failures, sensitivity should be assessed by varying the failure rate; it can be especially useful to use rates based on the placebo group rather than assuming failure for all patients with the ICE.

16.4.4 Sensitivity Analyses for Principal Stratification Strategies

Principal stratification approaches rely on covariates to estimate to which stratum each patient belongs when in fact the strata are not observable in a parallel arm trial (ICH, 2019). To ensure preservation of causality and unbiasedness, it is assumed that the probabilities of stratum memberships are correctly estimated by including all relevant factors in the model (Ratitch et al., 2019a). This, like missing data assumptions, cannot be tested from the data. Therefore, sensitivity to stratum membership should be investigated.

Such assessments could be made using alternative models to estimate stratum membership. Or stratum membership could be arbitrarily perturbed or naturally perturbed as part of a multiple imputation approach to assess how the estimated treatment effect varies. For details on sensitivity analyses in a principal stratification setting, see Shepherd, Gilbert, and Dupont (2011) and Schwartz, Li, and Reiter (2012).

16.4.5 Sensitivity Analyses for While-on-Treatment Strategies

While-on-treatment estimands typically require effects to be temporary and essentially unchanging over time, so that the timing of ICEs is not relevant. In practice, this means that while-on-treatment strategies are useful for outcomes that are summarized by a rate and/or hazard that is constant over time (Ratitch et al., 2019a). It is therefore important to assess sensitivity to this constancy assumption (e.g., linearity of change over time, proportionality of hazard).

16.4.6 Sensitivity Analyses for Treatment Policy Strategies

In treatment regimens that include more than just the initially randomized treatments, the actual regimens taken may vary across the sample. For instance, patients adhere to the initially randomized treatment, or progress onto another, of potentially many, allowed treatments. An assumption is therefore required that patients follow regimens in the trial in a similar, or at least relevant way, compared to those in the "real world" of clinical practice (Ratitch et al., 2019a).

If patients follow a sequence of treatments that is not plausible in clinical practice, sensitivity could be addressed by assuming those patients followed some hypothetical regimen. For example, if patients discontinue treatment but remain in the trial, it may be useful to conduct sensitivity analyses wherein values for these patients are imputed as if they switched to standard of care rather than took no treatment because in clinical practice these patients would not remain untreated.

Although the treatment policy approach will typically yield fewer missing values than hypothetical strategies, some missing data are inevitable, and sensitivity to missing data should be assessed as previously described.

16.5 Summary

Historically, emphasis on sensitivity analyses focused on missing data, especially sensitivity to departures from MAR. The delta-adjustment and reference-based imputation approaches are useful in this context. All strategies for handling ICEs require assumptions around missing data because it is always possible that some relevant data are missing or available data are irrelevant. Although this affects hypothetical strategies the most, it is also relevant for composite strategies where relevant data are missing when no ICE occurred as well as for the treatment policy strategy when, as is usually the case, it is not possible to achieve 100% follow-up.

In addition to sensitivity to missing data, composite strategies to handling ICEs should be assessed for sensitivity to values or ranks assigned to patients with relevant ICEs. In the while-on-treatment strategy, sensitivity to the assumption of constancy (linearity, proportional, or constant hazard) should be assessed. In the principal stratification approach, sensitivity to stratum membership should be investigated.

Section IV

Technical Details on Selected Analyses

Each chapter in Section IV provides technical details on an analytic approach that can be used as the main estimator or a sensitivity analysis consistent with one or more of the five strategies for dealing with intercurrent events as outlined in the ICH E9(R1) Addendum. An example data set is used, and the example code is provided to implement the analyses.

17

Example Data

17.1 Introduction

Throughout this section, technical details of selected analyses are illustrated using an example data set that includes 50 patients (25 per arm) with three post-baseline assessments. This data set is based on data from real clinical trial subjects but represents a subset of the actual trial-enrolled population. The following section provides more details on the data set and how it was created. Aspects of these data are admittedly arbitrary, and the intent is not to mimic any specific clinical trial setting but rather to provide a data set for illustration and convenience. A listing of these data is provided for those readers who wish to replicate and/or expand on the analyses presented here.

17.2 Details of Example Data Set

Two versions of the data set were created. The first version (complete data) had complete data where all patients adhered to the originally assigned study medication. The second version (missing data) was identical to the first except some data were missing such as would arise from patient dropout. Each data set had 50 subjects, 25 per arm, and 3 post-baseline assessments. A complete listing of these data sets is provided at the end of this chapter. The outcome data comprised of the HAMD17 (Hamilton 17-item rating scale for depression; Hamilton 1960) and Patient Global Impression of Improvement scale (PGIIMP; Guy 1976). The HAMD is a continuous variable. The PGI has seven ordered categories from "very much improved" (1) to "very much worse" (7), with "not improved" corresponding to the midpoint score =4.

17.2.1 Complete Data Set

The complete data set was created by extracting patients from a clinical trial in major depressive disorder (Detke et al., 2004). All subjects that completed the trial and were from the investigational site with the largest enrollment were

TABLE 17.1

Baseline Means by Treatment and Visitwise Means by Treatment in Complete Data

Trt	Time	N	Mean	Median	Deviation Standard
1	Baseline	25	19.80	20	3.06
1	1	25	−4.20	−4	3.66
1	2	25	−6.80	−6	4.25
1	3	25	−9.88	−10	4.85
2	Baseline	25	19.32	20	4.89
2	1	25	−5.24	−6	5.49
2	2	25	−8.60	−8	5.39
2	3	25	−13.24	−13	5.54

TABLE 17.2

Simple Correlations Between Baseline Values and Post-baseline Changes in Small Example Data Set

	Baseline	Time 1	Time 2	Time 3
Baseline	1.00	−0.26	−0.32	−0.03
Time 1		1.00	0.76	0.52
Time 2			1.00	0.71
Time 3				1.00

selected. Additional subjects, who were also completers, were selected to provide 25 subjects per arm. These additional subjects were selected to yield a data set that had a significant treatment effect at endpoint. Only data from weeks 2, 4, and 8 were included and then renamed as Time 1, Time 2, and Time 3.

Baseline and visitwise means by treatment are summarized for the complete data set in Table 17.1, and correlations between the baseline and post-baseline assessments are summarized in Table 17.2. Mean baseline values were similar across treatments. Post-baseline changes increased over time in both treatments (negative changes indicate improvement), with greater changes in Treatment 2. Baseline values had moderate-to-weak negative correlations with post-baseline changes such that subjects with greater symptom severity at baseline tended to have greater post-baseline improvement. Post-baseline values had strong positive correlations across time.

17.2.2 Data Set with Dropout

The data set with (monotone) dropout had 18 completers for Treatment 1, the "placebo" group, and 19 completers for Treatment 2, the "experimental" group. Observations were deleted from the complete data to mimic a realistic scenario in which subjects in the experimental arm were at greater risk of dropout due to adverse events. This mechanism was implemented by requiring a greater level of improvement in the experimental group than in the placebo group before subjects were "immune" to (had zero probability of) dropout. Consequently, the outcomes for subjects who discontinued from the placebo arm were worse than placebo subjects who remained in the study and worse than experimental group subjects who discontinued. The value that triggered the dropout was included in the data. The number of subjects by treatment and time is summarized in Table 17.3.

Visitwise means by treatment are summarized for the data set with dropout in Table 17.4.

Compared to the complete data set, the difference between treatments in mean change at Time 3 based on the observed data was somewhat smaller in the data with dropout. To further investigate the potential impact of incomplete data, the visitwise mean changes from baseline for completers versus dropouts are summarized in Figure 17.1. Subjects that discontinued early tended to have worse outcomes than those that completed the trial, mimicking dropout for lack of efficacy. However, Treatment 2 (drug-treated) subjects that discontinued after Time 1 tended to have better outcomes than those that completed the trial, thereby potentially mimicking dropout for adverse events.

TABLE 17.3

Number of Subjects by Treatment and Time in Small Data Set with Dropout

	Time		
Treatment	1	2	3
1	25	20	18
2	25	22	19
Total	50	42	37

TABLE 17.4

Visitwise Raw Means in Data with Dropout

Trt	TIME	N	Mean	Median	Standard Deviation
1	1	25	−4.20	−4.0	3.66
1	2	20	−6.80	−5.5	4.63
1	3	18	−10.17	−9.0	4.88
2	1	25	−5.24	−6.0	5.49
2	2	22	−8.14	−8.0	5.27
2	3	19	−13.11	−13.0	5.44

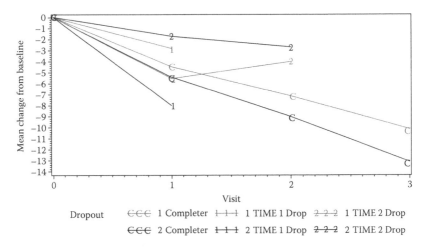

FIGURE 17.1
Visitwise mean changes from baseline by treatment group and time of last observation in the example data set with dropout.

Data from the small example data sets are listed in Tables 17.5 and 17.6.

TABLE 17.5

Listing of HAMD17 Data from Small Example Data Set

			Changes from Baseline					
			Complete Data			Data with Dropout		
Subject	Treatment	Baseline Value	Time 1	Time 2	Time 3	Time 1	Time 2	Time 3
1	2	24	−11	−16	−24	−11	.	.
2	1	20	−6	−8	−5	−6	.	.
3	2	18	−1	−1	−9	−1	−1	.
4	2	10	−9	−6	−9	−9	−6	−9
5	1	12	−6	−3	−9	−6	−3	−9
6	1	14	−6	−10	−10	−6	−10	−10
7	2	17	−7	−7	−14	−7	−7	−14
8	1	21	−2	−9	−9	−2	−9	−9
9	1	19	−9	−6	−10	−9	−6	.
10	2	19	−13	−12	−14	−13	−12	−14
11	2	20	−11	−15	−20	−11	−15	−20
12	2	19	−7	−15	−19	−7	.	.
13	1	20	−9	−12	−13	−9	−12	−13
14	1	19	−6	−12	−16	−6	−12	−16

(Continued)

TABLE 17.5 (*Continued*)

Listing of HAMD17 Data from Small Example Data Set

| | | | Changes from Baseline | | | | | |
| | | | Complete Data | | | Data with Dropout | | |
Subject	Treatment	Baseline Value	Time 1	Time 2	Time 3	Time 1	Time 2	Time 3
15	2	19	−12	−15	−18	−12	−15	−18
16	1	19	−3	−11	−17	−3	−11	−17
17	2	20	−9	−13	−19	−9	−13	−19
18	1	23	−7	−10	−15	−7	−10	−15
19	1	26	−5	−5	−11	−5	−5	−11
20	2	19	0	−1	−8	0	−1	−8
21	2	24	−12	−15	−19	−12	−15	−19
22	2	19	0	−2	−10	0	−2	−10
23	2	20	−7	−8	−13	−7	−8	−13
24	1	20	1	−1	−6	1	−1	−6
25	1	22	0	−4	−9	0	−4	−9
26	2	23	−12	−17	−22	−12	−17	−22
27	1	21	−1	−2	−3	−1	−2	−3
28	1	21	−2	−2	−2	−2	−2	−2
29	2	20	−3	−8	−13	−3	−8	−13
30	1	19	−2	−2	0	−2	−2	.
31	2	13	−1	−4	−11	−1	−4	−11
32	1	24	−10	−14	−20	−10	−14	−20
33	2	18	−4	−10	−15	−4	−10	−15
34	1	21	−2	−1	−6	−2	−1	−6
35	2	20	−5	−10	−15	−5	−10	−15
36	1	20	−4	−4	−10	−4	.	.
37	2	20	−3	−6	−12	−3	−6	.
38	1	22	−3	−5	−6	−3	−5	−6
39	2	20	−6	−9	−13	−6	−9	−13
40	1	18	−5	−9	−15	−5	.	.
41	2	20	5	−2	−10	5	−2	−10
42	1	15	0	−2	−8	0	−2	−8
43	1	19	−3	−9	−11	−3	.	.
44	2	8	−6	−3	−7	−6	−3	−7
45	1	20	−9	−11	−9	−9	−11	−9
46	2	20	−6	−5	−9	−6	.	.
47	2	35	−8	−14	1	−8	−14	1
48	1	17	−10	−14	−14	−10	−14	−14
49	1	23	4	−4	−13	4	.	.
50	2	18	−1	−1	−9	−1	−1	.

TABLE 17.6

Listing of PGI Improvement from the Example Data Set

Subject	Treatment	Time 1	Time 2	Time 3
1	2	3	2	2
2	1	4	4	5
3	2	4	4	4
4	2	6	2	1
5	1	2	5	4
6	1	2	2	2
7	2	2	2	2
8	1	4	2	2
9	1	2	3	2
10	2	2	2	1
11	2	3	2	2
12	2	2	2	1
13	1	3	3	2
14	1	3	1	1
15	2	3	3	2
16	1	3	2	2
17	2	3	2	2
18	1	3	3	2
19	1	4	4	3
20	2	4	4	3
21	2	3	3	1
22	2	4	3	2
23	2	3	3	2
24	1	4	4	2
25	1	4	3	2
26	2	2	2	1
27	1	4	4	4
28	1	4	4	4
29	2	4	3	2
30	1	4	4	4
31	2	2	2	1
32	1	2	2	1
33	2	4	3	2
34	1	4	4	3
35	2	3	2	2
36	1	3	3	2
37	2	3	3	2
38	1	3	3	4
39	2	2	2	2
40	1	2	1	1

(Continued)

TABLE 17.6 (*Continued*)

Listing of PGI Improvement from the Example Data Set

Subject	Treatment	Time 1	Time 2	Time 3
41	2	4	3	2
42	1	4	3	2
43	1	4	4	3
44	2	1	2	1
45	1	5	3	4
46	2	3	5	3
47	2	5	7	.
48	1	3	3	3
49	1	4	3	3
50	2	4	4	4

18

Direct Maximum Likelihood

18.1 Introduction

If missing data arises from a missing completely at random (MCAR) or missing at random (MAR) mechanism, a likelihood-based analysis can yield unbiased estimates and valid inferences from the observed data, such that the missing data can be ignored, provided the models are correctly specified. With generalized estimating equations (GEEs) and least squares estimation, for the missingness to be ignorable, it must generally arise from an MCAR mechanism, given their nonlikelihood basis. Therefore, the plausibility of ignorable missing data is greater with likelihood-based or Bayesian analyses than with GEE, least squares, or other frequentist methods (Verbeke and Molenberghs, 2000).

Section 18.2 covers technical details of likelihood-based estimation, and Section 18.3 provides the technical explanation for why MCAR and MAR missingness can be ignored in likelihood-based analyses. The examples in Section 18.4 illustrate how likelihood-based methods account for missing data.

18.2 Technical Details of Likelihood Estimation for Repeated Measures

Maximum likelihood (ML) estimation for normally distributed data using variance component models was considered by Crump (1947). The landmark papers on ML estimation include Hartley and Rao (1967), in which the first asymptotic results for the maximum likelihood estimators (MLEs) were established. Thompson (1962) introduced restricted maximum likelihood. Patterson and Thompson (1971) extended those ideas. Harville (1977) presented a comprehensive review of maximum likelihood and restricted maximum likelihood estimation in linear mixed-effects models. Laird and Ware (1982) and Jennrich and Schluchter (1986) are also standard references for likelihood estimation with mixed-effects models.

In general, a likelihood function describes how "plausible" (likely) a value for the parameter vector is, given the observed sample. In maximum likelihood estimation, the values of the parameter vector that maximize the likelihood function are chosen as the estimates for those parameters. For example, an appropriate plausible candidate for the likelihood function for a continuous variable may be the product of the normal probability densities associated with the observed values, which includes parameters for the mean and variance. A key implication is the need to estimate parameters for the mean and variance. When extending this to repeated measures taken in a longitudinal clinical trial, parameters for mean, variance, and covariance (correlation) need to be estimated.

With likelihood-based estimation and inference (as well as with Bayesian inference), missing data can be ignored if arising from either an MCAR or MAR mechanism and provided the technical condition of separability holds (i.e., the parameter space of the parameter describing the outcomes jointly with that describing the missingness mechanism is equal to the product of the individual parameter spaces). This is an extremely important advantage for longitudinal clinical trial data analyses over least squares and GEE that generally require the more restrictive assumption of MCAR for ignorability (in some rare exceptions, MAR would still yield valid frequentist inferences).

Use of least squares does not mean the errors (residuals from the model) will be small, only that no other estimates will yield smaller errors. Moreover, maximum likelihood does not guarantee a parameter estimate has a high likelihood of being the true value; only that there is no other value of the parameter that has a greater likelihood, given the data. In certain situations, maximum likelihood and least squares yield the same results. An example is complete, balanced outcomes modeled with a linear regression with normally distributed errors.

Restricted maximum likelihood estimates (RMLEs) are sometimes preferred to MLE in linear mixed-effects models because RMLEs consider estimation of fixed effects when calculating the degrees of freedom associated with the variance-component estimates, while MLEs do not. Therefore, RMLEs reduce and in some cases eliminate small-sample bias present in MLE. As Verbeke and Molenberghs (2000) describe, this comes at the cost of no RMLE-based likelihood ratio test for fixed effects, but other tests (e.g., Wald tests) are still available and valid.

The maximum likelihood and RMLEs are obtained by constructing appropriate objective functions and maximizing that function over the unknown parameters. The corresponding objective functions are as follows:

$$\text{ML:} \quad l(\mathbf{G}, \mathbf{R}) = -\frac{1}{2}\log|\mathbf{V}| - \frac{1}{2}\mathbf{r}'\mathbf{V}^{-1}\mathbf{r} - \frac{n}{2}\log(2\pi)$$

$$\text{REML:} \quad l_R(\mathbf{G}, \mathbf{R}) = -\frac{1}{2}\log|\mathbf{V}| - \frac{1}{2}\log|\mathbf{X}'\mathbf{V}^{-1}\mathbf{X}| - \frac{1}{2}\mathbf{r}'\mathbf{V}^{-1}\mathbf{r} - \frac{n-p}{2}\log(2\pi)$$

where $\mathbf{r} = \mathbf{y} - \mathbf{X}(\mathbf{X}'\mathbf{V}^{-1}\mathbf{X})^{-}\mathbf{X}'\mathbf{V}^{-1}\mathbf{y}$ and p is the rank of $\mathbf{X}$.

In practice, commercial software packages minimize twice the negative of these functions by using, for example, a ridge-stabilized Newton–Raphson algorithm or other quasi-Newton methods, which are generally preferred over the expectation-maximization (EM) algorithm (Lindstrom and Bates, 1988) because of speed of convergence. One advantage of using the Newton–Raphson algorithm is that the second derivative matrix of the objective function evaluated at the optima (H) is available upon completion. The same is true for Fisher scoring, with the expected rather than the observed information matrix available.

The asymptotic theory of maximum likelihood shows that H^{-1} is an estimate of the asymptotic variance-covariance matrix of the estimated parameters of G and R. Therefore, tests and confidence intervals based on asymptotic normality can be obtained. However, these tests can be unreliable in small samples, especially for parameters such as variance components that have sampling distributions that tend to be skewed to the right (SAS, 2013). For more details, see Verbeke and Molenberghs (2000) on estimation and testing of variance components in likelihood-based models.

18.3 Factoring the Likelihood Function for Ignorability

Factorization of the likelihood function in the missing data context means that the hypothetical "full" data consists of the observed and the missing part (governed by the measurement process) and a missingness indicator (governed by the missingness process) (Verbeke and Molenberghs, 2000).

As a building block to factoring the full data likelihood, Mallinckrodt and Lipkovich (2017) provided the following: consider the observed data for subject i, denoted as a $1 \times n_i$ vector $y_{i.obs}$ and the observed $1 \times n$ vector of missingness indicators R_i that indicate whether the subject has the outcome observed on jth occasion ($r_{ij} = 1$) or missing ($r_{ij} = 0$). The observed data likelihood can be written as the joint density of the random variables $Y_{i.obs}$ and R_i, $f(y_{i.obs}, r_i \mid x_i, \theta, \Psi)$, where X is the design matrix for fixed effects (e.g., baseline covariates, treatment, time), θ is a vector of parameters for the outcome process, and Ψ is a vector of parameters for the missingness process.

The aim in clinical trials is typically to estimate θ, with Ψ being of ancillary interest. As such, it would be desirable to partition the joint (log-) likelihood of $(Y_{i.obs}, R_i)$ to isolate the portion of the likelihood that is relevant for estimating θ. With this partitioning, the parts of the likelihood associated with the missingness parameters are not relevant for maximizing the likelihood with respect to θ – and these missingness parameters can therefore be ignored.

The joint likelihood written as the product of marginal and conditional distributions is

$$f\left(y_{i.obs}, r_i \mid x_i, \theta, \Psi\right) = f\left(y_{i.obs} \mid x_i, \theta, \Psi\right) f\left(r_i \mid y_{i.obs}, x_i, \theta, \Psi\right)$$

or on the log-likelihood scale as the sum:

$$\log\left[f\left(y_{i.obs}, r_i \mid x_i, \theta, \Psi\right)\right] = \log\left[f\left(y_{i.obs} \mid x_i, \theta, \Psi\right)\right] + \log\left[f\left(r_i \mid y_{i.obs}, x_i, \theta, \Psi\right)\right].$$

In this decomposition, θ and Ψ are present in both pieces of the likelihood in the right-hand part, and thus Ψ is not ignorable. Fortunately, if missingness results from an MCAR or MAR mechanism and the parameters θ and Ψ are functionally separable (the so-called separability condition, see also the previous section), then a partitioning to isolate the parameters of interest is attainable.

To see how this is possible, consider the full data likelihood that includes both the observed and the missing data. Let a random variable $Y_{i.mis}$ denote the $1 \times (n - n_i)$ component vector of missing data for the ith subject. Although $Y_{i.mis}$ represents data that is not observed, it is a valid random variable representing the potential outcomes for subjects who discontinued from the trial that would have been observed had they remained.

The observed data likelihood can be written as the integral over the $Y_{i.mis}$. Under the integral sign, the joint density is as follows:

$$f\left(y_{i.obs}, r_i \mid x_i, \theta, \Psi\right) = \int f\left(y_{i.obs}, y_{i.mis}, r_i \mid x_i, \theta, \Psi\right) dy_{i,mis}$$

$$= \int f\left(y_{i.obs}, y_{i.mis} \mid x_i, \theta\right) f(r_i \mid y_{i.obs}, y_{i.mis}, x_i, \Psi) dy_{i,mis}$$

With the introduction of Y_{mis}, the joint likelihood for the full data can be factored into pieces that separate the parameters θ and Ψ because of the assumption that the complete outcome $(Y_{i,obs}, Y_{i.mis})$ does not depend on the missingness parameter Ψ, and the missingness process (R) does not depend on θ *after conditioning on the complete* data.

Under MAR, $f(r_i \mid y_{i.obs}, y_{i.mis}, x_i, \Psi) = f(r_i \mid y_{i.obs}, x_i, \Psi)$. Therefore, this likelihood factor can be taken outside the integral, and the desired factorization of the observed data likelihood is obtained as follows:

$$f\left(y_{i.obs}, r \mid x_i, \theta, \Psi\right) = f\left(y_{i.obs} \mid x_i, \theta\right) f(r_i \mid y_{i.obs}, x_i, \Psi).$$

Now, parameter θ is associated only with the likelihood term containing $y_{i,obs}$. Therefore, estimation of θ (and any functions of treatment effect of our interest) requires only contributions to the observed data likelihood of $y_{i.obs}$, $i = 1,..., N$ because the parts associated with parameters of missingness (Ψ) are not relevant for maximizing the likelihood with respect to θ.

This elegant proof (see Little and Rubin, 1987) is general and applies to any type of outcome for which a likelihood can be written. The proof may be appealing to mathematically inclined readers. However, the accounting for missing data in the context of longitudinal clinical trials is not intuitively obvious in this analytic framework that was initially focused on the sample survey context. A more operational and instructive explanation in the longitudinal clinical trial context of how direct likelihood methods and, in particular, methods based on an unstructured modeling of time and covariance yield unbiased estimates under MAR can be given by a factorization of the likelihood for the observed part of the full repeated measures outcome $Y_i = (Y_{i1}, Y_{i2}, \ldots, Y_{in})$.

Consider a small example data set with three post-baseline outcomes, represented for the ith subject by the trivariate normal variate $Y_i = (Y_{i1}, Y_{i2}, Y_{i3})$ and further assume that all the missingness is due to dropout (i.e., the missing data has a monotone pattern). Focus first for simplicity on analysis of a single treatment arm because adding covariates does not change things in principle. Thus, all time-fixed covariates are denoted collectively as X.

The trivariate normal distribution associated with random variate Y_i can be factored as a product of the conditional distributions that represent a sequence of regressions, where each component of the distribution is regressed on the previous outcomes and baseline scores. Subject indices (i) are suppressed for simplicity.

$$f(y|x,\theta) = f(y_1|x,\theta_1) f(y_2|x,y_1,\theta_2) f(y_3|x,y_1,y_2,\theta_3)$$

Parameter θ combines the vector of means and covariance matrix of multivariate normal distribution of Y conditional on X and parameters θ_1, θ_2, and θ_3 contain regression coefficients and error variances associated with the three univariate conditional normal distributions (regressions): $[Y_1 | X]$ $[Y_2 | X, Y_1]$, and $[Y_3 | X, Y_1, Y_2]$. Multivariate normal distributions have a one-to-one relation between parameters in θ and in $(\theta_1, \theta_2, \theta_3)$, so θ can be obtained by simple matrix manipulations from θ_1, θ_2, and θ_3 and vice versa.

Consider obtaining the maximum likelihood estimates of θ via sequentially fitting the regression models above, even though that is not how the likelihood maximization procedure is implemented in software packages. If the parameters θ_1, θ_2, and θ_3 estimated by the three regression models are unbiased under MAR, then θ would also be unbiased. The key question is why are the regression models based on observed data unbiased and not adversely affected by the presence of missing data?

It is important to recall that the MAR assumption for monotone missingness means that the statistical properties of the future (counterfactual) outcomes for subjects who discontinued are similar to those of subjects who continued *if they had the same values for previous outcomes, covariates, etc.* Therefore, complete data is not mandatory for estimating any of the three regression models. For example, the model based on fitting $f(y_{3,obs}|x,y_{1,obs},y_{2,obs},\theta_3)$ to observed data should estimate the same parameter vector θ_3 as if somehow the model was fit to the missing data: $f(y_{3,mis}|x,y_{1,obs},y_{2,obs},\theta_3)$. Therefore, failing to observe (and include) $y_{3,mis}$ does not bias estimates of θ_3.

Although the concept of MAR is appealing, it is important to recognize that the validity of MAR can never be proven. Also, there is another important consideration wherein an MAR analysis may perform poorly even when MAR is valid. The MAR assumption can be represented in terms of potential outcomes, so that the counterfactual outcomes for subjects who discontinued at time t are from the same distribution as subjects with the same outcome and covariate history up to time t who remained on treatment. However, if no subjects with a similar history up to time t remained on trial, direct likelihood analyses rely on extrapolation. Therefore, even though the mechanism is MAR, the analysis may perform poorly.

Such situations occur whenever rules for mandatory discontinuation for lack of efficacy are in place, or if the target estimand requires data after rescue to be censored, rules for mandatory use of rescue medication would have the same effect. In these situations, the imputation model for multiple imputation is poorly informed, and inverse probability weighting analysis may not even be possible because of the required assumption that for all outcome histories, the probability of observing future outcomes must be >0.

18.4 Example

The small example data set introduced in Chapter 17 is used to illustrate how direct likelihood analyses account for missing data. Results are compared to the corresponding complete data set. Selected individual subjects are used for further illustration. Data can be analyzed in SAS or R as specified in Code Fragment 18.1. This model features an unstructured modeling of means and covariance. (See additional details on model fitting for repeated measures data in Chapter 9). Results are summarized from the data with dropout and the corresponding complete data is provided in Table 18.1.

TABLE 18.1

Results from Likelihood-Based Analyses of Complete and Incomplete Data, with a Model Including Baseline as a Covariate

Treatment	Time	Complete Data		Incomplete Data	
		SE	LSMEANS	SE	LSMEANS
1	1	−4.13	0.91	−4.10	0.91
1	2	−6.70	0.93	−6.42	0.97
1	3	−9.86	1.05	−9.73	1.17
2	1	−5.32	0.91	−5.29	0.91
2	2	−8.70	0.93	−8.52	0.96
2	3	−13.26	1.05	−12.62	1.14
Endpoint Treatment Difference		3.39	1.49 ($p = .0274$)	2.90	1.64 ($p = .084$)

Results at Time 1 were similar in complete and incomplete data. Differences in results between complete and incomplete data were greater at Time 3 than at Time 2. This pattern was expected given no data were missing at Time 1 and that more data were missing at Time 3 than at Time 2.

As the amount of missing data increased, standard errors increased. Visit-wise LSMEANS for change from baseline were smaller for both treatment groups in the incomplete data than in the complete data. This disparity was greater in Treatment 2, thereby yielding a smaller treatment contrast from incomplete data than from complete data.

It is important to interpret the results above as just one realization from a stochastic process. Replicating the comparisons many times under the same conditions – with missing data arising from an MAR mechanism – would yield equal averages of the LSMEANS and treatment contrasts for complete and incomplete data. Standard errors would be consistently greater in incomplete data, however.

To further clarify how missing data is handled in likelihood-based analyses, consider the observed and predicted values for selected subjects summarized in Table 18.2. First, consider the incomplete data for drug-treated

TABLE 18.2

Observed and Predicted Values for Selected Subjects from Analyses of Complete and Incomplete Data

Time	Observed Change[a]	Group Mean	Pred Change	StdErr Pred	Residual[a]
Subject 1 Complete Data					
1	−11	−5.32	−12.36	2.16	1.36
2	−16	−8.70	−16.12	2.18	0.12
3	−24	−13.26	−19.20	2.27	−4.79
Subject 1 Incomplete Data					
1	−11	−5.29	−10.03	2.70	−0.96
2	.	−8.52	−13.49	3.34	.
3	.	−12.62	−15.39	4.58	.
Subject 30 Complete Data					
1	−2	−4.13	0.20	2.17	−2.20
2	−2	−6.70	−2.32	2.17	0.32
3	0	−9.86	−5.67	2.23	5.67
Subject 30 Incomplete Data					
1	−2	−4.10	−1.34	2.51	−0.65
2	−2	−6.42	−3.62	2.52	1.62
3	.	−9.73	−6.34	3.85	.

[a] In the observed change column, "." indicates missing data. In the residual column, "." indicates that a residual could not be calculated because the corresponding outcome was missing.

(Treatment 2) Subject 1 who dropped from the study after the first post-base-line visit. Although Time 2 and 3 were not observed, the parameters for the estimated treatment mean factor-in the general dependency of Y_2 and Y_3 on Y_1 – which can also be used to generate predicted outcomes. This is because under MAR the future values (Y_2 and Y_3) for Subject 1, and other subjects who discontinued after Time 1, follow the relationships associated with the conditional distributions of $[Y_2 | Y_1]$ and $[Y_3 | Y_1, Y_2]$.

Predicted values for Subject 1 suggest that the unobserved outcomes at Times 2 and 3 would have shown continued improvement, and the improvement for Subject 1 would have been greater than the group average. Factors influencing this prediction include (1) the group to which Subject 1 belonged showed continued improvement over time and (2) the positive correlation between the repeated measurements suggested that subjects (such as Subject 1) with outcomes better (worse) than their group mean were likely to remain better (worse) than the group mean in the future.

Recall that first having complete data and then deleting observations according to an MAR mechanism created the example data. Therefore, it is useful to compare predicted outcomes from the incomplete data to the observed complete data and to predicted values obtained from the complete data. Predicted values for Subject 1 in the complete data deviate more from the group mean than in the incomplete data. For example, at Time 3, the predicted value for Subject 1 is approximately six points above the mean in complete data and only three points above the mean in incomplete data. With incomplete data, there is less evidence for the superior improvement of Subject 1; therefore, the above average improvement seen at Time 1 leads to a predicted value at Time 3 that is regressed (shrunk) more strongly back to the group mean than when data were complete. The standard errors of the predicted values increase substantially when the corresponding observation is missing.

Similar relationships exist for Subject 30, which came from the placebo (Treatment 1) group. However, this subject had smaller improvements than the group mean. In the incomplete data, the predicted values for Subject 30 were less than the group average, reflecting the anticipation of continued below average performance. As with Subject 1, incomplete data resulted in greater shrinkage of predicted values back to the group mean than in complete data. This increased shrinkage with incomplete data is again due to the evidence for the deviation in performance being weaker with fewer observations. However, Subject 30 had two observations, and the shrinkage in predictions in the incomplete data was less than if only one value was observed. Correspondingly, the standard errors of the predicted values are smaller for Subject 30 than for Subject 1.

18.5 Code Fragments

Code Fragment 18.1 SAS and R code for fitting a model with unstructured time and unstructured within-patient error covariance

SAS code

```
PROC MIXED DATA=ONE;
  CLASS SUBJECT TRT TIME;
  MODEL CHANGE = BASVAL TRT TIME BASVAL*TIME TRT*TIME /
    DDFM=KR;
  REPEATED TIME / SUBJECT=SUBJECT TYPE =UN R;
  LSMEANS TRT*TIME/DIFFS;
  ODS OUTPUT DIFFS=_DIFFS (where=(TIME= _TIME))
    LSMEANS=_LSMEANS ;RUN;
```

R code

```
require(nlme)
require(contrast)

fitgls.erun <- gls(change ~ basval +trt+ TIME +
  basval*TIME + trt*TIME, data = complete,
weights = varIdent(form= ~ 1 | TIME),
correlation=corSymm(form=~ as.numeric(TIME)| subject))

summary(fitgls.erun)

# computing error variance-covariance matrix
getVarCov(fitgls.erun)

# evaluating least squares means at time 3
contrast(fitgls.erun ,a = list(trt = "1", basval=
  mean(complete$basval), TIME = "3"),type = "individual")

contrast(fitgls.erun ,a = list(trt = "2", basval
  = mean(complete$basval), TIME = "3"),type = "individual")

# evaluating treatment contrast at time 3
contrast(fitgls.erun ,a = list(trt = "1",
  basval  =mean(complete$basval), TIME = "3"),
  b = list(trt = "2",basval= mean(complete$basval),
  TIME = "3"),type  = "individual")
```

18.6 Summary

If missing data arises from an MCAR or MAR mechanism, a likelihood-based analysis yields unbiased estimates and valid inferences from the available data such that the missing data process can be ignored. Factorization of the full data likelihood into parts that isolate the missingness parameters from the parameters underlying the outcome process provides the technical explanation for why MCAR and MAR missingness can be ignored in likelihood-based analyses.

A more intuitive explanation for ignorability stems from the definition of MAR. Recall the following two ways to consider MAR: (1) The probability of missingness does not depend on missing data given observed data and (2) the distribution of unobserved future outcomes is the same as the distribution of observed future outcomes, conditional on earlier outcomes. The second condition means that the future outcomes for subjects who discontinued should be similar to future outcomes of subjects who continued *if they had the same values of past (observed) outcomes, covariates, etc*. Because of (2), models formulated from only the observed data yield unbiased estimates of parameters describing the full data.

19

Multiple Imputation

19.1 Introduction

Multiple imputation (MI) is a popular and accessible method of model-based imputation. Standard references for MI include Rubin (1987) and Little and Rubin (2002). MI is flexible and therefore has several specific implementations. The three basic steps to MI are as follows:

- Impute the missing data (typically) using Bayesian predictive distributions of missing data, conditional on observed data, resulting in multiple (m) completed data sets.

- Analyze the m completed data sets using an analysis that would have been appropriate for complete data. This results in a set of m estimates (e.g., treatment contrasts).

- Combine (pool) the m estimates into a single inferential statement using combination rules (or "Rubin's rules") that account for uncertainty due to imputation of the missing values, therefore providing valid inference.

Applications of MI include continuous outcomes that follow a normal distribution, categorical outcomes, recurrent events (count data), or time-to-event (e.g., exponentially distributed) outcomes where missing data results from censoring. MI can also be applied when outcomes with different data types (e.g., continuous, discrete, count) are simultaneously analyzed, for example, as part of a composite clinical score.

Implementing MI under the missing at random (MAR) assumption with the same imputation and analysis model, that is the same as the analysis model used in a maximum likelihood (ML)-based analysis, yields asymptotically similar point estimates for the two approaches. That is, as the size of the data set and the number of imputations in MI increase, estimates from MI and ML converge to a common value. Although ML is somewhat more efficient (smaller standard errors) (Wang and Robins, 1998), MI is more flexible. The distinct steps for imputation and analysis in MI yield flexibility that, as explained in subsequent sections, can be exploited in several situations. A down side to

MI is that the three-step process must be applied for each parameter of interest. Alternatively, multiple parameters can be handled simultaneously using a multiparameter version of the combination rules in step 3, with the additional burden of estimating the variance covariance matrix associated with estimated parameters, even though many packages would provide these by default.

19.2 Technical Details

Point estimates in MI are computed as a simple average of the estimates from the m completed data sets. The precision of these estimates is evaluated using formulas that incorporate both between-imputation and within-imputation variability in the calculation of standard errors. In the following, we provide technical details assuming, for simplicity, the test statistics and associated estimates are univariate; however, the formulas readily generalize for multivariate test statistics.

Let $\hat{\theta}_i, i = 1,...,m$ be univariate estimates (e.g., treatment contrasts at the last scheduled visit) obtained by applying an appropriate analysis model to the m completed data set and $U_i, i = 1,...,m$ are their estimated (squared) standard errors. The MI point estimate is

$$\hat{\theta}_{MI} = \frac{\sum_{i=1}^{m} \hat{\theta}_i}{m}.$$

The total variance is computed by combining the between-imputation variability, $B = (m-1)^{-1} \sum_{i=1}^{m} (\hat{\theta}_i - \hat{\theta}_{MI})^2$, and the within-imputation variability, $W = m^{-1} \sum_{i=1}^{m} U_i$,

$$V = W + \left(1 + \frac{1}{m}\right) B.$$

Inference on θ is based on the test statistic $T = \frac{\hat{\theta}_{MI}}{\sqrt{V}}$. As with ML, the $\hat{\theta}_{MI}$ estimator is unbiased if missing data arises from a missing completely at random (MCAR) or missing at random (MAR) mechanism.

The null distribution of T is not normal. Therefore, standard Wald-based inference assuming T has a standard normal null distribution is not valid because for finite $m < \infty$ the total variance V is an inconsistent estimate of the true variance of the point estimator, $var(\hat{\theta}_{MI})$. That is, when the sample size n increases to infinity, nV does not converge to a constant but remains a random variable following a nondegenerate (chi-squared) distribution. This contrasts with, for example, the variance of a sample mean: $N[\widehat{var}(\bar{x})] = \frac{N}{N}\hat{\sigma}^2 \xrightarrow{p} \sigma^2 = const.$

To account for this extra variability, Rubin (1987) proposed using a Student's t distribution instead of a normal distribution, with the number of degrees of freedom computed as

$$v = (m-1)\left(1 + \frac{1}{1+m^{-1}}\frac{W}{B}\right)^2.$$

The price paid for the generality of the MI estimator is a degree of conservatism (sometimes negligible). This overestimation of variability has three sources (Wang and Robins, 1998; Robins and Wang, 2000).

- The asymptotic variance of the MI point estimator (for finite m) exceeds that of the ML estimator
- The standard error of the MI point estimator obtained by Rubin's rules may exceed its true value (i.e., overestimation of the standard error)
- Inconsistency of Rubin's variance estimator requiring use of the wider confidence intervals based on the t distribution rather than the normal distribution

Each missing value is imputed *multiple* times, which allows for explicit accounting for uncertainty due to the imputation/missing data by incorporating the between-imputation variability. In MI, imputations are simulated from the Bayesian posterior predictive distribution of missing data given observed data $f(y_{mis}|y_{obs})$ rather than from the density $f(y_{mis}|y_{obs},\hat{\theta})$ with parameters θ estimated at specific values (e.g., by ML). The predictive distribution is obtained by integrating (averaging) out parameters from the likelihood using the posterior distribution of parameters.

$$f\left(y_{mis}|y_{obs}\right) = \int f\left(y_{mis}|y_{obs},\theta\right)f(\theta|y_{obs})d\theta.$$

Therefore, imputed values are sampled from distributions that incorporate uncertainty in estimating model parameters and uncertainty due to sampling data from the estimated model. As a simple example, when estimating the mean and variance of a univariate normal distribution $N(\theta,\sigma^2)$, MI accounts for uncertainty in the estimated θ and σ^2 as well as uncertainty of sampling missing values from $Y_{mis} \sim N\left(\hat{\theta},\hat{\sigma}^2\right)$.

It is instructive to think of imputing (sampling) from a predictive distribution as a two-stage procedure that is repeated m times. Repeat the following two steps m times for $i = 1,...,m$:

- Draw a single random sample from the posterior distribution of parameters given observed data $\tilde{\theta}_i \sim f(\theta \mid y_{obs})$
- Draw random samples from the observed-data likelihood model $Y_{mis} \sim f(y \mid \tilde{\theta}_i)$. to produce a single completed data set

It is useful to consider the consequences of creating m completed data sets using only the second step (i.e., without re-sampling from the posterior of θ). In this approach, point estimates would still be valid, essentially reproducing the ML estimates. However, the total variance V computed using Rubin's rules would be too small because the between-imputation variability would not include the uncertainty due to parameter estimation. Consequently, the confidence intervals based on V would be too narrow, failing to maintain the nominal coverage rates. Such imputation models have been termed *improper* (Rubin, 1987).

With complex models, analytic formulas for the posterior distribution $f(\theta \mid y_{obs})$ often do not exist and Markov Chain Monte Carlo (MCMC) methods can be used (Tanner and Wong, 1987; Schafer, 1997). In the context of repeated measures analysis with a multivariate normal model, when the pattern of missingness is arbitrary, generic MCMC sampling must be used for imputation. When the pattern is monotone, sampling can be based on factorization of the joint normal distribution into a sequence of conditionals and using analytical formulas for the posterior distribution of parameters associated with each factor.

For the small example data set with dropout, the factorization of the likelihood is

$$f(y \mid x,\theta) = f(y_1 \mid x,\theta_1) f(y_2 \mid x,y_1,\theta_2) f(y_3 \mid x,y_1,y_2,\theta_3).$$

Then imputation can be organized sequentially

- Draw $\tilde{\theta}_1 \sim f(\theta_1 \mid x,y_1)$ and impute missing values for Y_1, $Y_{mis} \sim f(y_1 \mid x,\tilde{\theta}_1)$. (In the example data set, no y_1 are missing, and this step is not needed.)
- Draw $\tilde{\theta}_2 \sim f(\theta_2 \mid x,y_1,y_2)$ and impute missing Y_2 based on observed and imputed Y_1 with $Y_{mis} \sim f(y_2 \mid x,\tilde{y}_1,\tilde{\theta}_2)$. (Here $\tilde{y}_1$ indicates that Y_1 could be observed or imputed.)
- Draw $\tilde{\theta}_3 \sim f(\theta_3 \mid x,y_1,y_2,y_3)$ and impute missing Y_3 based on the observed and imputed Y_1 and Y_2, $Y_{mis} \sim f(y_3 \mid x,\tilde{y}_1,\tilde{y}_2,\tilde{\theta}_3)$.

Therefore, imputation of missing values is done via estimated Bayesian regression models where parameters are drawn from posterior distributions.

In normal linear regression, missing values for Y_2 (Time 2) are imputed by sampling values from the regression model $Y_{mis} = \tilde{\beta}_0 + \tilde{\beta}_1 X + \tilde{\beta}_2 \tilde{Y}_1 + e\tilde{\sigma}$, where X is the baseline score, $\tilde{Y}_1$ is observed or imputed value of the change from baseline at Time 1 and $\tilde{\beta}_0, \tilde{\beta}_1, \tilde{\beta}_2, \tilde{\sigma}^2$ are drawn from their respective posterior distributions. Random error e is sampled from the standard normal distribution, $e \sim N(0,1)$.

As an illustration, Figure 19.1 depicts the observed data and the means from 100 imputations using the above sequential Bayesian regression for subjects 1, 30, and 49 from the small example data set with dropout. The SAS and R code for this analysis is listed in Section 19.8 (Code Fragment 19.1). Results for treatment groups are discussed later. The focus here is on the individual subjects. In Figure 19.1, the error bars represent the between-imputation variability (standard deviation based on the 100 imputed values at each time point).

The trajectory for means of the imputed values resembles the conditional means based on a direct likelihood analysis as depicted in Table 18.2. Subjects with observed changes that were less than the average of their group have mean imputed values that are less than the corresponding group mean; subjects with observed changes that were greater than the group average have mean imputed values that are greater than the corresponding group means. Also, like the direct-likelihood analysis, the standard deviation bars indicate greater variability in imputed values at Time 3 for Subjects 1 and 49 who had only one observed value (at Time 1) compared with Subject 30 who had two observed values.

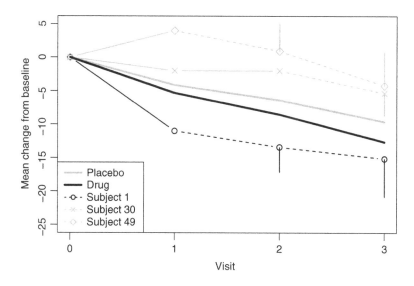

FIGURE 19.1
Illustration of multiply imputed values for Subjects 1, 30, 49 from the small example data set with dropout.

Thin solid lines depict the observed data, and dotted lines depict means of the imputed values for the focus subjects. Thick solid lines depict the treatment group means. The error bars indicate standard deviation of imputed values.

The concordance between direct-likelihood and MI results is not coincidental; these similarities are expected in all circumstances when the methods are implemented using similar models. MI can also be compared with direct Bayesian inference. Rubin (1987) showed that MI is compatible with Bayesian analysis in the sense that the MI point estimate and variance are approximating the posterior expectation and posterior variance of a Bayesian analysis.

In that regard, MI is somewhat circular. First, the Bayesian posterior distribution of parameters is sampled. Next, the distribution of missing values is simulated, and in the end, the posterior sample of parameters is discarded. Hence, the posterior distribution of parameters serves merely as an intermediate step in imputation. Alternatively, sampling from posterior distribution could continue and be used for purely Bayesian inference about parameters based on the output from posterior distribution of θs (and any of their functions).

As noted earlier in this section, when implemented with similar models, MI and ML yield asymptotically similar point estimates, but MI has greater variance and is therefore less efficient, meaning that ML would be somewhat more powerful. However, MI has other advantages, such as when some covariate values are missing. In likelihood-based analyses, if a covariate is missing for a subject, all post-baseline data for that subject are discarded unless much more complicated analyses are implemented where covariates are considered part of the outcome process as well. In MI, the missing covariates can easily be imputed, thereby preserving inclusion of the post-baseline data. An example of imputing missing covariates is provided in Section 19.5.

Other advantages of MI stem from having separate steps for imputation and analysis. This flexibility does not exist in ML because ML has only the analysis step, thereby implying that only the analysis variables can be used to account for the missing data.

In MI, the imputation model can include additional covariates not present in the analysis model. These covariates may be predictive of outcomes and/or the probability of dropout. Meng (1994) and Collins, Schafer, and Kam (2001) provide examples of and rationale for these so-called inclusive modeling strategies. An example of inclusive modeling is provided in Section 19.5.

The consequences of an analysis model that is richer (has more variables) than the imputation model is very different. Consider an imputation model that does not include treatment. The imputation model assumes no *direct* treatment effect and can capture the *indirect* treatment effect that may be mediated via intermediate outcomes (if included in the model). Consequently, imputed outcomes across treatment arms are likely to be more similar than should be the case. Analyses based on such imputed data sets therefore would be biased whenever a direct treatment effect existed.

Although the model in the example above is misspecified – in the context of MAR – it can also be thought of as a specific form of missing not at random (MNAR). In the above example, the imputed values systematically deviate from unbiased MAR estimates in a manner that decreases the magnitude of the estimated treatment contrast. This points to what is becoming an important usage of MI – as a convenient tool for conducting sensitivity analyses. The imputation models can be set up to generate specific departures from MAR, thereby fostering a sensitivity analysis for an MAR-based primary analysis. These reference-based and delta-adjustment implementations of MI are considered in Chapters 22 and 23, respectively.

19.3 Example – Implementing MI

19.3.1 Introduction

This section provides details on implementing several MI strategies using readily available statistical software applied to data from the small example data set with dropout. Focus is on relevant SAS procedures that are commonly used in practice; however, some examples using R are also provided.

In SAS, MI can be implemented via the following steps:

- Use PROC MI to generate m completed data sets
- Analyze each data set, resulting in m point estimates and standard errors
- Pass the point estimates and standard errors from step 2 to PROC MIANALYZE that implements Rubin's rules for combining results

The output from PROC MIANALYZE includes the upper and lower limits of the confidence interval and associated p-value for the two-sided null hypothesis that the parameter is at the specified value (e.g., $\theta = 0$). The parameter of interest for MI inference can be multivariate although the current examples are scalar parameters of treatment effects at specific time points.

The procedure outlined above is easy to implement and is a common approach in conducting MI analysis in SAS. However, variations on the approach are possible and sometimes needed. For example, the first step can be divided into two steps: (1) sampling from posterior distributions of parameters estimated by an appropriate likelihood-based model using a SAS procedure (different from PROC MI) and (2) given each posterior draw, impute missing values from the imputation model with custom programming code. An example could be imputing missing counts (such as in the analysis of recurrent events) from a Bayesian Poisson regression, which can be fitted separately using SAS PROC GENMOD with the BAYES statement. In some situations (e.g., when modeling growth curves in repeated measures data), multiple draws from posterior predictive distributions can be conveniently obtained via SAS PROC MCMC (available from SAS Version 9.3).

In some cases (particularly, for nonparametric and semiparametric models), sampling from posterior distributions can be replaced with nonparametric bootstrapping. For example, data can be sampled with replacement, resulting in m bootstrap samples from the original data. Posterior draws $\tilde{\theta}_1, \tilde{\theta}_2, \ldots, \tilde{\theta}_m$ are mimicked with parameter estimates from bootstrap samples $\hat{\theta}_1^*, \ldots, \hat{\theta}_m^*$. The imputed values are generated from conditional distributions, $Y_{mis} \sim f\left(y \mid \hat{\theta}_i^*\right), i = 1, \ldots, m.$

As an example, consider imputing time to event for a subject censored at time t_{cens}. Imputations can be based on Kaplan–Meier estimates of a survival distribution computed from m bootstrap samples: $\tilde{T} \sim \hat{S}_i^*\left(t \mid t > t_{cens}\right), i = 1, \ldots, m.$ Bootstrapping is often used in conjunction with hot-deck procedures. Single hot-deck imputation is generally not valid because it does not account for uncertainty due to missing data. However, using an approximate Bayesian bootstrap (such as the "Monotone Propensity Score Method" implemented in SAS PROC MI) allows constructing more valid hot-deck procedures. Values can be resampled from "donors" within propensity score groups that have similar probability of missingness. In this case, the bootstrap mimics sampling from predictive posterior distribution of missing data.

Using bootstrap to mimic posterior sampling of parameters – or the predictive distribution of missing data – was termed "approximate Bayesian bootstrap" in Little and Rubin (1987). This should not be confused with standard uses of bootstrapping as a method for obtaining standard errors and confidence intervals (e.g., via percentile or ABC methods). For example, 95% confidence intervals for the treatment contrast θ can be constructed in the familiar manner of using the 2.5%–97.5% percentiles of the bootstrap distribution of the MI point estimator $\hat{\theta}_{MI}$ as the confidence interval bounds.

Bootstrap confidence intervals are more computationally intensive than the usual confidence intervals based on Rubin's rules but can be useful when the latter have questionable validity. A natural approach for using the bootstrap for confidence intervals would be to carry out MI on each bootstrap sample and the MI point estimate computed from the m completed data sets. Other schemes can be entertained by first generating m completed data sets based on MI and then bootstrapping each of the completed sets. Schomaker and Heumann (2016) conducted simulations comparing several alternative bootstrap-based estimators.

Code Fragment 19.1 in Section 19.8 provides example SAS code to implement MI for the small example data set with dropout. The most commonly used R package for MI is *mice*. Example R code to implement MI is provided in Code Fragment 19.2.

19.3.2 Imputation

To use PROC MI, the data should be prepared in the so-called "multivariate" (or "wide" or "horizontal") format with data from each time point in a separate column. Longitudinal outcomes are typically stored as a "stacked" (or "long") data set, which is the format anticipated by SAS PROC MIXED. Data can be converted from the stacked format to the wide format using SAS PROC TRANSPOSE. The transposed data set is passed to PROC MI, and missing values are imputed using a multivariate normal model with treatment fitted as a covariate or imputations done separately by treatment.

The example data set has a monotone missingness pattern. Therefore, a Bayesian regression for multivariate normally distributed data can be used, thus avoiding the MCMC methods that are needed for imputing data with arbitrary missingness patterns. Checking whether the pattern is monotone can be conveniently done by running PROC MI with parameter NIMPUTE=0. Table 19.1 shows default output from PROC MI illustrating the missing data patterns by the treatment group. The output makes it easy to see that the response patterns in the example data are indeed monotone. If a subject has a missing value at a given visit, all subsequent observations are missing.

The order of variables listed in the VAR statement of PROC MI should correspond to the order of imputations. For the example data, the order of variables corresponds to the assessment times. Alternatively, the sequential orientation can be explicit using multiple statements. The code for explicit specification is in Code Fragment 19.1. The explicit specification is not mandatory for this example but provides flexibility that is useful in other situations illustrated in subsequent examples.

In this example, imputation is by treatment arm. This is equivalent to a single imputation model with treatment, treatment-by-visit and all treatment-by-covariate interactions. However, having separate imputation models by

TABLE 19.1

Missing Data Patterns for the Small Example Data Set with Dropout

		Variables[a]					
Trt	Group	Basval	Yobs1	Yobs2	Yobs3	Freq	Percent
	1	X	X	X	X	18	72.0
1	2	X	X	X		2	8.0
1	3	X	X			5	20.0
2	1	X	X	X	X	19	76.0
2	2	X	X	X		3	12.0
2	3	X	X			3	12.0

[a] "X" indicates presence of the outcome score at baseline (Basval) and 3 post-baseline visits (Yobs1, Yobs2, Yobs3), "." indicates the value is missing. "Groups" identifies the distinct patterns present in the data set.

treatment arm is simple and yields the same result as the more parameter rich single, saturated imputation model (with the minor difference that separate covariance structures would be fitted when using by-treatment processing, whereas a single pooled error covariance matrix will be used with the single imputation model).

The number of imputations (specified by the parameter NIMPUTE) chosen for this example is 1000. In the early MI literature, it was recommended that as few as $m = 5$ imputations was sufficient to achieve good precision of MI estimates. The basis for this recommendation was the relative efficiency of MI estimates compared to ML estimates. However, the decrease in power due to small m may substantially exceed that expected from the theoretical argument based on the relative efficiency (Graham et al., 2007). With modern computing power, it is feasible to generate hundreds and even thousands of imputations in reasonable time. The case for using a small number of imputations is obsolete unless in extremely large sets of data. For a summary of recent literature on choosing the number of imputations, see van Buuren (2018, Section 2.8, pp. 58–60).

As a practical tool for selecting a reasonable number of imputations, plot MI point estimates against the number of imputations, as done in Figure 19.2 for the small example data set with dropout. This plot allows visual assessment of the stability of MI estimator $\hat{\theta}_m$ for smaller m and its

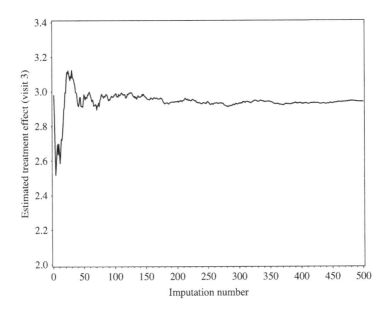

FIGURE 19.2
MI estimator $\hat{\theta}_m$ for the treatment contrast at visit 3 computed over the first m completed data sets versus the number of imputations (m).

"convergence" to $\hat{\theta}_\infty$. In Figure 19.2, after $m_0 = 200$, the point estimator is stable, which ensures the *reproducibility* of MI analyses for this data set with $m \geq m_0$. The sufficient number of imputations varies from one data set to another, with a general trend for larger data sets and data sets with lower rates of missing data to require fewer imputations to stabilize. In all examples for this Chapter $m = 1000$ although this may be larger than needed.

In PROC MI, the output data set has the same "multivariate" structure as the input data set. However, the now completed multiple data sets are *stacked*, and the variable named _Imputation_ (assuming values 1,...,<*nimpute*>) is automatically generated to indicate to which data set each observation belongs. Figure 19.3 shows a fragment of the imputed (completed) data set. Imputed values in this example have a larger number of decimal points. In some applications, auxiliary variables (with values 0 and 1) are useful to distinguish observed and imputed values. This requires writing custom SAS code before calling PROC MI.

	Imputation	trt	subject	basval	Yobs1	Yobs2	Yobs3
1	1 1		2	20	-6	-5.371893093	-10.27327664
2	1 1		5	12	-6	-3	-9
3	1 1		6	14	-6	-10	-10
4	1 1		8	21	-2	-9	-9
5	1 1		9	19	-9	-6	-12.78127318
6	1 1		13	20	-9	-12	-13
7	1 1		14	19	-6	-12	-16
8	1 1		16	19	-3	-11	-17
9	1 1		18	23	-7	-10	-15
10	1 1		19	26	-5	-5	-11
11	1 1		24	20	1	-1	-6
12	1 1		25	22	0	-4	-9
13	1 1		27	21	-1	-2	-3
14	1 1		28	21	-2	-2	-2
15	1 1		30	19	-2	-2	-6.425250822
16	1 1		32	24	-10	-14	-20
17	1 1		34	21	-2	-1	-6
18	1 1		36	20	-4	-7.120718604	-14.13814589
19	1 1		38	22	-3	-5	-6
20	1 1		40	18	-5	-6.057750783	-2.045101917
21	1 1		42	15	0	-2	-8
22	1 1		43	19	-3	-6.167970307	-10.26103763
23	1 1		45	20	-9	-11	-9
24	1 1		48	17	-10	-14	-14
25	1 1		49	23	4	4.4830432444	2.5435936604
26	2 1		2	20	-6	-11.29465031	-11.65938348
27	2 1		5	12	-6	-3	-9
28	2 1		6	14	-6	-10	-10
29	2 1		8	21	-2	-9	-9
30	2 1		9	19	-9	-6	-16.37321523
31	2 1		13	20	-9	-12	-13
32	2 1		14	19	-6	-12	-16

FIGURE 19.3
Fragment of complete data set produced by PROC MI.

19.3.3 Analysis

After completing the imputation step, an appropriate analysis model can be applied to each completed set. For longitudinal data, as in the example data set, it is natural to consider applying the same direct likelihood analysis of repeated measures via PROC MXIED to each completed data set as for the analysis of incomplete data. While this analysis approach is useful for assessing the longitudinal response profiles, it is also possible to implement a simple visitwise analysis of variance (ANOVA) or analysis of covariance (ANCOVA) if interest is only on the cross-sectional contrasts. This is possible because imputation fully accounts for the missing data (if assumptions hold true), and there is no benefit from further accounting for the missing data via a repeated measures analysis.

When applying the analysis model to imputed data, it is useful to take advantage of SAS's "BY-processing" capability using the "by _Imputation_" statement in the analysis procedure. If a repeated measures analysis of the imputed data is desired, an additional data manipulation step is needed to transpose the PROC MI output data set back to the "stacked" format required with repeated measures data in PROC MIXED. In this example, the simple ANCOVA model is implemented to analyze the Time 3 variable (Yobs3 in the imputed data set). Figure 19.4 illustrates the point estimates and standard errors from the analyses of some of the imputed data sets.

	Imputation	Label	Estimate	StdErr	DF	tValue	Probt	time
267	67	trt effect at vis 3	-1.7813	1.4729	47	-1.21	0.2325	3
268	68	trt effect at vis 3	-3.4297	1.4091	47	-2.43	0.0188	3
269	69	trt effect at vis 3	-3.1104	1.5551	47	-2.00	0.0513	3
270	70	trt effect at vis 3	-2.3183	1.5762	47	-1.47	0.1480	3
271	71	trt effect at vis 3	-2.5356	1.4717	47	-1.72	0.0915	3
272	72	trt effect at vis 3	-4.4898	1.6927	47	-2.65	0.0109	3
273	73	trt effect at vis 3	-2.9977	1.4591	47	-2.05	0.0455	3
274	74	trt effect at vis 3	-3.4374	1.5221	47	-2.26	0.0286	3
275	75	trt effect at vis 3	-1.8695	1.7225	47	-1.09	0.2833	3
276	76	trt effect at vis 3	-3.6741	1.5684	47	-2.34	0.0234	3
277	77	trt effect at vis 3	-3.5779	1.4950	47	-2.39	0.0207	3
278	78	trt effect at vis 3	-3.8548	1.5870	47	-2.43	0.0190	3
279	79	trt effect at vis 3	-3.5465	1.4240	47	-2.49	0.0163	3
280	80	trt effect at vis 3	-4.0039	1.7441	47	-2.30	0.0262	3
281	81	trt effect at vis 3	-4.0162	1.4955	47	-2.69	0.0100	3
282	82	trt effect at vis 3	-4.0411	1.3845	47	-2.92	0.0054	3
283	83	trt effect at vis 3	-3.5701	1.4470	47	-2.47	0.0173	3
284	84	trt effect at vis 3	-2.3854	1.3850	47	-1.72	0.0916	3

FIGURE 19.4
Fragment of results from the analyses of multiply imputed data sets to be used as input for PROC MIANALYZE.

19.3.4 Inference

The last step in MI is to combine estimates and standard errors to obtain a single inferential statement. Code Fragment 19.3 shows the SAS code for PROC MIANALYZE to accomplish this step. The "BY TIME" statement is used because the input data set contains treatment contrasts for Time 2 and Time 3. Note that Time 1 did not have missing values; therefore all estimates for the treatment contrast at Time 1 obtained from the multiply imputed data set are identical, and, of course, it does not make sense to run PROC MIANALYZE for Time 1. The results are shown in Table 19.2. Note that the input data set for PROC MIANALYZE should contain a mandatory variable _Imputation_ that indicates for each estimate and associated standard error the number of the imputation data set from which it was computed.

The LSMEANS, treatment contrasts, and standard errors from the MI analysis of the incomplete data are, as expected, similar to the corresponding results from direct likelihood (see Table 18.1). Compared with the complete data, the MI results show a smaller treatment contrast, larger standard errors, and a larger *p*-value for the endpoint treatment contrast. It is important to interpret these results as just one realization from a stochastic process. Replicating the comparisons many times under the same conditions – with missing data arising from an MAR mechanism – would yield average LSMEANS and treatment contrasts across the repeated trials that were approximately equal for complete and incomplete data. Standard errors would be consistently greater in incomplete data, however.

TABLE 19.2

Treatment Contrasts and Least-Squares Means Estimated by Multiple Imputation from the Small Example Data Set with Dropout

Treatment	Time	Complete Data		Incomplete Data (MI), $m = 1000$	
		LSMEANS	SE	LSMEANS	SE
1	2	−6.70	0.93	−6.36	1.01
1	3	−9.86	1.05	−9.61	1.19
2	2	−8.70	0.93	−8.50	0.99
2	3	−13.26	1.05	−12.56	1.23
Endpoint Treatment Difference		3.39	1.49 ($p = 0.0274$)	2.95	1.73 ($p = 0.0971$)

19.3.5 Accounting for Non-monotone Missingness

Only monotone missingness has been addressed thus far. PROC MI with the MCMC statement can accommodate non-monotone missingness patterns. To illustrate these procedures, intermittent missingness was introduced into the example data by deleting some outcomes at Time 2 for subjects that had no other missing data. Time 2 outcomes were deleted for three subjects in the treated group (2) and for two subjects in the placebo group (1).

The top panel (a) of Code Fragment 19.4 (Section 19.8) provides an example of PROC MI code for imputation from a multivariate normal distribution using MCMC. Missing values are imputed from a single chain of the MCMC algorithm. Recall that MCMC samples converge to the target posterior distribution as the number of samples becomes large. Therefore, initial samples are skipped to ensure that the draws occur sufficiently close to the target distribution. This skipping is specified in the number of burn-in iterations (NBITER=200). Also because of serial correlation in sampled values within the same chain, a thinning period is specified as the number of iterations between imputations in a chain. In the example NITER=100 and PROC MI skip 100 samples after each imputation. The values NBITER=200 and NITER=100 are defaults that would be used if these options are omitted but are explicitly specified here for illustration.

It is useful to inspect the various MCMC diagnostics (e.g., autocorrelation plots); imputations should be rerun with increased values of NBITER and NITER if issues exist with MCMC convergence (see PROC MI help for more details). The example code explicitly specified that the initial values for the MCMC chain are posterior modes of parameters obtained by the EM (expectation-maximization) algorithm, with the maximum number of iterations set at MAXITER=200. Increase this number if the EM algorithm fails to converge within the specified limit. Nonconvergence is more likely in data sets with larger fractions of missing data or if there are sparse patterns of missing values. If convergence was not achieved, the SAS log file includes a warning.

The MCMC approach may be rather cumbersome. It (typically) requires more computing time compared with the Bayesian regression methods for monotone missingness. The additional layer of approximation inherent in MCMC requires monitoring for convergence to the desired target distributions. In contrast, the regression methods for monotone data sample *directly* from the target distributions and so there is no worry about convergence.

Often, only a few intermittently missing values cause the patterns to deviate from monotone because a few patients missed assessments while remaining in the study. In these situations, it is efficient to first impute the intermittent missing values using the MCMC option in PROC MI to complete the pattern to monotone; then, a second invocation of PROC MI can implement methods for monotone missingness, such as the Bayesian regression from the previous example. This is a situation in which fully conditional specification (FCS) works well (see the next section for additional details on FCS).

The SAS code for implementing this two-stage imputation strategy is in the bottom panel (b) of Code Fragment 19.4 (Section 19.8). Two runs of PROC MI are required. The first run produces partially completed data sets, each having a monotone pattern of missing values. The second run takes the output data set from the first run as input, processed by _Imputation_variable, and completes each data set with a single imputation from Bayesian regression for monotone data.

19.4 Situations Where MI Is Particularly Useful

19.4.1 Introduction

This section outlines situations when MI-based analyses provide certain advantages compared to direct likelihood (ML) or present a reasonable analytic strategy in absence of appropriate direct likelihood methods. Subsequent sections provide detailed examples.

19.4.2 Scenarios Where Direct Likelihood Methods Are Difficult to Implement or Not Available

Multiple imputation can be particularly useful in situations where direct likelihood methods are hard to implement or not available. Missing baseline covariates make direct likelihood methods (e.g., via SAS PROC MIXED) suboptimal because subjects with a missing covariate are not used for analysis even if they have complete post-baseline data. In contrast, MI can impute baseline covariates jointly with outcome variables (see Section 19.5 for example), thereby utilizing the records with missing covariates.

Repeated categorical outcomes are hard to analyze using direct likelihood methods because – in contrast to multivariate normal distributions for repeated continuous outcomes – multivariate distributions for binary or count data do not have simple and natural formulations. Imputing missing categorical data is relatively easy, especially when sequential imputation strategies can be applied, such as when missingness is monotone.

For outcomes of mixed types (continuous and categorical), MI can be implemented either by sequential imputation (for monotone patterns of missing data) or in the case of arbitrary missingness by repeatedly sampling from full conditional distributions (van Buuren, 2007). The method of FCS is implemented in the R package *mice* (multivariate imputation by chained equations) and recently in SAS PROC MI (via FCS statement available from SAS Version 9.3 and later).

Unlike joint modeling, FCS specifies the multivariate distributions through a sequence of conditional densities of each variable given the other variables. With FCS, it is even possible to specify models for which no joint distribution (of a posited form) exists – a situation referred to as incompatibility. Although incompatibility is not desirable, in practice it is a relatively minor problem, especially if the missing data rate is modest (van Buuren, 2018. Section 4.5.3, pp. 122–124).

19.4.3 Exploiting Separate Steps for Imputation and Analysis

Other situations where MI is particularly useful are those in which the separation of the imputation and analysis stages can improve efficiency (compared to direct likelihood methods).

Recall that the original motivation for MI in Rubin (1978a) was for analysis of complex surveys where the *imputer* had access to more information than the *analyst*. The idea was that after imputing missing data using the richer set of data, completed data sets with a limited number of variables would be available for analysts. Several clinical trial situations are considered where the imputation model can be based on more information than is included in the analysis model.

Sometimes clinical endpoints specified in the primary analysis are *derived* outcomes based on the underlying outcomes. In such situations, an efficient MI strategy could include four steps:

- Impute missing data for the underlying outcome measures
- Compute derived outcomes from the completed data sets
- Analyze completed data
- Combine (pool) results for inference.

An example of a derived variable is the analysis of "clinical response," defined as x% improvement in some continuous scale (or a combination of several scales). Imputing the underlying continuous measures may improve efficiency compared to analysis (or imputation and analysis) of the binary outcome.

As noted in Section 19.6, MI can utilize auxiliary variables in the imputation model that help predict the outcome and/or the dropout process. These auxiliary variables may not be available or are undesirable for use in the analysis model (Meng, 1994; Collins, Schafer, Kam, 2001). For example, the analysis model should not include intermediate outcomes or *mediators* of treatment effect, which would result in diluting the estimated treatment effect in a likelihood-based analysis. Classic examples of such variables would be time-varying post-baseline covariates that are influenced by treatment, such as secondary efficacy outcomes.

The following situations are examples of inclusive imputation strategies that can be useful.

- Pretreatment covariates (demographic and illness characteristics) and perhaps their interactions with treatment) are included in the imputation model. The selection of covariates can be based on their ability in predicting the outcome or in predicting the dropout. This can be evaluated in separate modeling steps, and a subset of covariates can be selected from a broader set of candidate covariates using appropriate model selection techniques such as stepwise selection, LASSO, or other machine learning methods.

- Post-baseline time-varying covariates. Inclusive imputation models can *jointly* impute two related outcomes, such as two efficacy scales, Y1 and Y2; e.g., two clinical rating scales that reflect different aspects of the disease. "Borrowing" information from Y2 may improve imputation of Y1. Section 19.6 includes an example where joint imputations of the scores for clinician-rated Hamilton Depression Scale (HAMD) and a patient-reported outcome scale (Patient Global Impression of Improvement [PGIIMP]) are conducted. The basic idea is that incorporating both the clinician's and patient's perspective on response may help explain subsequent HAMD outcomes and/or better inform the missingness process.

- When the outcome of interest is collected at sparse visit intervals or only at the last scheduled evaluation, whereas other outcomes are observed more frequently. Examples include efficacy and safety scales collected at scheduled time points and spontaneously recorded outcomes (occurrences of hospitalizations, adverse events). As another example, imputation of a binary response defined at a late stage of treatment can be imputed jointly with early signs of response represented by continuous scales. The binary outcome can be derived from a continuous scale (examples in Ratitch et al., 2018) or it can be inherently binary, based on clinical assessments or objective events (see Lipkovich and Wiens, 2018).

An approach that at first may seem counterintuitive but can be useful is to use safety assessments to help impute efficacy outcomes. Suppose that dropouts are predicted by both previous outcomes and future (unobserved outcomes) so the missingness is MNAR. Then MI using only earlier outcomes is inadequate and may result in biased estimates. However, suppose that conditional on both recent safety assessments (occurrence of specific adverse events) and earlier efficacy outcomes the dropouts *do not* depend on future efficacy outcomes. This may be the case when AEs are early signs/predictors of emerging changes (sudden worsening, say) in the outcome that is not predicted from the efficacy outcome alone. In this scenario, adding safety assessments in the imputation model may help satisfy the MAR assumption.

19.4.4 Sensitivity Analysis

The flexibility of separate analysis and imputation steps can facilitate creation of relevant departures from MAR, thereby facilitating a sensitivity analysis for an MAR primary analysis. The same imputation approaches can be thought of as modeling plausible outcomes after treatment discontinuation or switching treatment in hypothetical approaches to handling intercurrent events (ICEs). Implementation of these MNAR analyses involves controlling the imputation procedures or by manipulating imputed outcomes prior to data analysis (see Chapter 22 for details and examples).

It is important to appreciate the point raised earlier in this chapter that when implemented in similar manners, MI and ML have similar assumptions and yield similar results. Therefore, MI implemented similarly to ML is not a sensitivity analysis. However, controlling the imputations creates relevant departures from MAR and facilitates the sensitivity assessments or different hypothetical strategies to handling ICEs.

19.5 Example – Using MI to Impute Covariates

19.5.1 Introduction

In the small example data set with missing values, there were no missing baseline values. In well-conducted clinical trials, missing data in baseline covariates are usually minimal. However, occasionally some subjects may have missing baseline values. Moreover, analysis models may include stratification covariates or other covariates that could be missing. In such scenarios, the chance that at least one covariate is missing for a given subject can be appreciable (say 5%–10%). In these situations, typical likelihood-based analyses are not advisable because subjects with missing covariates are excluded from the analysis, resulting in loss of efficiency and potentially biased estimates.

19.5.2 Implementation

MI provides several strategies for imputing missing covariates. An obvious route is to use a two-stage strategy similar to that demonstrated in Section 19.3 for data with non-monotone missingness. First, use MCMC-based imputation to impute missing covariates. Then impute the missing post-baseline values to create the complete data sets. Code Fragment 19.5 (Section 19.8) illustrates the SAS code for imputing covariates and missing post-baseline covariates in a scenario with baseline severity and three covariates (X1, X2, and X3). The approach is similar to that in Code Fragment 19.4, thereby illustrating that only minor modification is required to impute missing covariates.

Another potential strategy may be to impute only missing baseline covariates using MCMC (PROC MI) and then applying a direct likelihood repeated measures analysis to multiple data sets with imputed baseline covariates and using Rubin's combination rules as we would normally do when imputing all missing values. Comparing results from this MI approach to the analysis that had to exclude all subjects with missing covariates isolates the impact of missing baseline covariates on the estimated treatment effect and associated standard errors. Imputing missing covariates should improve efficiency and lead to smaller standard errors when compared to deleting records with missing covariates. Moreover, if baseline covariates are not missing completely at random, estimates of the treatment effect may be biased if subjects with missing covariates are excluded.

To illustrate the efficiency of multiply imputing covariates alone, artificial missingness in baseline severity was created in the small example data set with dropout by deleting the baseline scores of five subjects who had unfavorable outcomes in HAMD changes at the last scheduled visit. Least-square means and treatment contrasts from a direct likelihood analysis (SAS PROC MIXED) of the 45 subjects with nonmissing baseline scores in the small example data set with dropout are compared to results from MI of the missing covariates in Table 19.3. The left-side columns are results from the direct likelihood analysis that excluded subjects with missing covariates. The right-side columns are results based on the same repeated measures model fitted to data sets with baseline scores imputed using all available outcome variables (via PROC MI, MCMC statement with the IMPUTE=MONOTONE option). As mentioned, normally both baseline and post-baseline values would be imputed. However, the approach used here is valid under MAR and allows explicit assessment of the impact of having to exclude subjects with missing covariates in the likelihood-based analysis.

The results with MI of the missing covariates, as expected, had smaller standard errors. Excluding the five patients with missing baseline scores also resulted in approximately a one-half point increase in the endpoint treatment

TABLE 19.3

Treatment Contrasts and Least-Squares Means with and without Imputation of Missing Covariates in the Small Example Data Set with Dropout

Treatment	Time	Covariates Missing[a]		Covariates Imputed[b]	
		LSMEANS	SE	LSMEANS	SE
1	1	−4.08	0.90	−4.21	0.94
1	2	−6.55	1.05	−6.56	1.00
1	3	−9.79	1.21	−9.78	1.12
2	1	−6.05	0.92	−5.25	0.94
2	2	−9.05	1.05	−8.45	0.98
2	3	−13.36	1.21	−12.79	1.09
Endpoint Treatment Difference		3.56	1.71 ($p = 0.0443$)	3.01	1.56 ($p = 0.0606$)

[a] Direct likelihood analysis with five subjects having missing covariates who are therefore excluded from the analysis.
[b] Direct likelihood analysis after multiple imputation of missing baseline covariates.

contrast. Of course, this is just one contrived example from a small example data set. It is not possible to know how including versus not including subjects with missing covariates will affect point estimates in actual clinical trial scenarios. However, regardless of scenario, including subjects with missing covariates increases the plausibility of MAR and reduces the standard errors compared with excluding those subjects.

This contrived example also helps to illustrate why imputation models for covariates should include post-baseline outcomes. Including post-baseline outcomes for imputing covariates may seem counterintuitive in that the model is predicting the past from the future. However, the approach can be justified theoretically and is supported by simulation studies (Crowe, Lipkovich, and Wang, 2009).

The imputation model does not have to be a causal model. In fact, if the missingness mechanism is such that subjects with poorer outcomes are more likely to have missing baseline scores, failing to include future outcomes when imputing baseline covariates may result in biased estimates of the fixed effects associated with these covariates.

Whether this bias would translate into biased estimates of treatment contrasts is not a clear-cut issue. While including future outcomes in imputation model may protect against aforementioned bias, it may induce another type of bias. Note that in an RCT, treatment assignment is independent of baseline covariates due to randomization, in which case bias in estimates of covariate effects would not bias treatment contrasts. However, an incorrect imputation model (provided it includes post-baseline measures) may induce correlation between the treatment variable and covariates. Treatment contrasts could be biased indirectly through the biased estimates of covariates (see also Sullivan et al., 2018).

In general, the imputation model should be compatible with the analysis model. That is, if an effect is included in the analysis model it should be in the imputation model.

19.6 Examples – Using Inclusive Models in MI

19.6.1 Introduction

Section 15.5 introduced and contrasted inclusive and restrictive modeling frameworks. This section provides an example of an inclusive imputation strategy for the small example data set with dropouts. In this example, as elsewhere in this book, the primary outcome of interest is change in the HAMD scale. The difference here is that the PGIIMP scale is used in addition to HAMD scores to "improve" the imputation of the missing HAMD scores.

Recall from the descriptions in Chapter 17 that the HAMD is a clinician-rated symptom severity score that focuses explicitly on specific symptoms and combines responses on the 17 individual items into a total score. The PGIIMP scale is a patient-rated measure of global improvement taken at each post-baseline visit. The PGIIMP assesses patient's overall feeling since they started taking medication and ranges from 1 (Very much better) to 7 (Very much worse). The missingness mechanism in this small example data set is such that if a HAMD value is missing, the corresponding PGIIMP value is also missing.

The clinical and statistical justification for this inclusive strategy stems from the PGIIMP reflecting *patients'* overall perspective, which may capture aspects not fully captured by the *clinician*-rated HAMD scale.

19.6.2 Implementation

The first step in jointly imputing HAMD and PGIIMP using SAS PROC MI is to transpose data for both outcomes from the "stacked" to the "horizontal" format. That is, the output data set has three columns for changes in HAMD (CHG1, CHG2, CHG3) and three columns for PGIIMP1, PGIIMP2, PGIIMP3 that correspond to post-baseline Times 1, 2, and 3. Columns are also included for baseline severity on the HAMD and treatment group. There are no baseline assessments for PGIIMP because it is inherently a measure of change.

The next step is to check for monotonicity of missingness patterns on *both* outcomes. For this purpose, the following order of the outcome variables is assumed: CHG1 PGIIMP1, CHG2 PGIIMP2, and CHG3 PGIIMP3. Table 19.4 displays the missing data patterns summarized by SAS PROC MI that indicate the joint patterns are monotone. One patient with a HAMD score at Time 3 had a missing Time 3 PGI-Improvement (Group 2, Treatment 2 in Table 19.4). The pattern is still monotone because PGIIMP3 goes after CHG3 in the ordering of variables. Imputation of values for PGIIMP3 is not needed because focus is on HAMD and PGIIMP3 does not inform imputation for CHG3. Hence, the PGIIMP3 column could be omitted with no impact on HAMD results.

Consider two imputation approaches. First, treat all outcomes as continuous and use sequential Bayesian regression for multivariate normal data to impute both HAMD changes and PGI-Improvement. In this approach, it is possible to impute missing values for PGIIMP that are outside the scales range of 1–7. Therefore, the validity of applying a normal distribution to categorical PGIIMP data is questionable. Schafer (1997) showed that imputation is robust to some deviations from multivariate normality. Also, Lipkovich et al. (2014) provide simulation evidence in favor of using a multivariate normal distribution for imputing such categorical data.

The second approach avoids assuming multivariate normality for PGI-Improvement. The process is to sequentially implement: (1) linear regression to impute changes in HAMD conditional on earlier HAMD changes and on PGI-Improvement and (2) ordinal logistic regression to impute missing PGIIMP values conditional on earlier (observed or imputed) HAMD changes and PGIIMP scores.

TABLE 19.4

Missingness Patterns for Joint Modeling of Changes in HAMD and PGI-Improvement

				Missing Data Patterns					
Trt=1 (Placebo)									
Group	Basval	CHG1	PGIIMP1	CHG2	PGIIMP2	CHG3	PGIIMP3	Frequency	Percent
1	X	X	X	X	X	X	X	18	72.00
2	X	X	X	X	X	.	.	2	8.00
3	X	X	X	.	.	.	.	5	20.00
Trt=2 (Treatment)									
Group	basval	CHG1	PGIIMP1	CHG2	PGIIMP2	CHG3	PGIIMP3	Frequency	Percent
1	X	X	X	X	X	X	X	18	72.00
2	X	X	X	X	X	X	.	1	4.00
3	X	X	X	X	X	.	.	3	12.00
4	X	X	X	.	.	.	.	3	12.00

Implementing this second approach in SAS PROC MI requires specifying PGIIMP1 and PGIIMP2 as class variables in both linear and logistic regression models. Recall that a PGIIMP score of 4 corresponds to "no change," and scores > 4 indicate worsening. In these data, PGIIMP scores > 4 are uncommon. Therefore, all scores > 4 were combined with scores of 4 (i.e., recoded as "level 4") and the rescaled PGI scores for Time 1 and Time 2 were named PGIIMPGR1 and PGIIMPGR2, respectively.

The SAS code to implement the imputation strategy using a multivariate normal is provided in the top panel (a) of Code Fragment 19.6 (Section 19.8) and the code to implement the normal-ordinal model is in the bottom panel (b). In PROC MI, it is possible to sample from truncated marginal distributions when imputing PGIIMP2 scores, thereby ensuring the scores are within the scale range 1–7. However, this approach was shown to be inferior to nontruncated imputation (Lipkovich et al., 2014). Therefore, truncation is not used here.

Imputation of categorical outcomes by an ordinal logistic model is similar in spirit to imputation by linear regression (explained in Section 19.2). Code Fragment 19.6 implements Bayesian ordinal logistic regression fitted separately for each treatment arm to predict PGIIMP scores at Time 2 given PGIIMPGR1 (class variable), CHG1 (continuous variable), and baseline HAMD score (continuous variable). Then, for a single draw from the posterior distribution of model parameters, the probability for each of the four categories: P1, P2, P3, P4 (adding up to 1) is estimated, and imputed values are generated as a multinomial random variable.

For example, the probability of category "2" is computed as the difference Prob(PGIIMP2="2")=Prob(PGIIMP2≤"2")−Prob(PGIIMP2≤"1"), whereas Prob(PGIIMP2≤"2") and Prob(PGIIMP2≤"1") are directly estimated from the assumed logistic model.

$$\text{logit}\left[\text{Prob}\left(Y_i \leq l_k\right)\right] = a_k + \sum x_{ij}\beta_j,$$

where Y_i denotes the PGI Improvement score for the ith patient, l_k, $k = 1,2,3$ correspond to the categories "1," "2," "3" of the outcome variable, and x_{ij} represents the observed or imputed score for the jth covariate. The time dimension and specific covariate structure (including for this case BASVAL and CHG1) are suppressed to simplify notation.

Results from the two models are summarized in Table 19.5. The treatment contrasts from the normal and logistic-normal imputation models were 2.69 and 2.87, respectively. Recall the treatment contrast from "standard" MI in Table 19.2 was 2.95. It is important to interpret these results as an illustration. The intent is to show the potential usefulness of, and the process for, inclusive MI models. Results from this small example data set do not inform expectation of results for actual clinical trial scenarios.

The Code Fragment 19.7 illustrates using the R package *mice* for multiply imputing continuous outcomes jointly with categorical PGIIMP scores.

TABLE 19.5

Treatment Contrasts and Least-Squares Means Estimated by Multiple Imputation: Changes in HAMD Using Joint Model for HAMD and PGIIMP

Treatment	Time	Incomplete Data (MI) (Joint Multivariate Normal Model)		Incomplete Data (MI) (Joint Modeling Using Normal Model for HAMD and Ordinal Logistic for PGIIMP)	
		LSMEANS	SE	LSMEANS	SE
1	2	−6.37	1.02	−6.35	1.01
1	3	−9.72	1.22	−9.76	1.38
2	2	−8.45	0.99	−8.59	1.00
2	3	−12.41	1.20	−12.63	1.30
Endpoint Treatment Difference		2.69	1.71 ($p = 0.1248$)	2.87	1.90 ($p = 0.1393$)

19.7 MI for Categorical Outcomes

As explained in Chapter 9, likelihood-based analyses can be difficult to implement for categorical outcomes. Therefore, MI provides an important framework for MAR-based analyses of categorical data. MI can be used for analysis of binary outcomes, including those derived from an underlying continuous scale and for an ordinal categorical variable. For inherently binary data, a logistic model is used for imputation. When the binary outcome is dichotomized from an underlying continuous variable, it is preferable to impute the missing continuous outcome and subsequently dichotomize the imputed outcome.

19.8 Code Fragments

Code Fragment 19.1   SAS Code for MI analysis. Creating completed data sets with proc MI using monotone imputation

```
/* transpose original data resulting in WIDE data with outcome
columns Yobs1 Yobs2 Yobs3 corresponding to post-baseline
visits 1,2,3 */
  PROC SORT DATA=ALL2; BY  TRT SUBJECT BASVAL; RUN;
  PROC TRANSPOSE DATA=ALL2 OUT= ALL2_TRANSP2 (DROP=_NAME_)
  PREFIX=YOBS;
    BY TRT SUBJECT BASVAL;
    VAR CHGDROP;
    ID  TIME;
  RUN;
```

```
/* check whether pattern is monotone */
  PROC MI DATA = ALL2_TRANSP2 NIMPUTE =0;
   BY TRT;
   VAR BASVAL YOBS1 YOBS2 YOBS3;
  RUN;

/* perform multiple imputation for monotone data */
  PROC MI DATA = ALL2_TRANSP2 OUT = ALL2_MIOUT NIMPUTE=1000
  SEED=123;
   BY TRT;
   MONOTONE METHOD=REG;
   VAR BASVAL YOBS1 YOBS2 YOBS3;
  RUN;

/* multiple imputation of monotone data with explicit model
statement*/
 PROC MI DATA = ALL2_TRANSP2 OUT = ALL2_MIOUT_2 NIMPUTE=1000
 SEED=123;
   BY TRT;
   VAR BASVAL YOBS1 YOBS2 YOBS3;
   MONOTONE REG (YOBS1 =BASVAL);
   MONOTONE REG (YOBS2 =YOBS1 BASVAL);
   MONOTONE REG (YOBS3 =YOBS2 YOBS1 BASVAL);
  RUN;
```

Code Fragment 19.2   Example R Code for MI analysis of continuous outcome with arbitrary missingness: change from baseline on HAMD

```
require(mice)

# assumes transposed data set in multivariate format
head(hamdchng)
  trt basval  CHG1  CHG2 CHG3
1   1     20    -6    NA   NA
2   1     12    -6    -3   -9
3   1     14    -6   -10  -10
4   1     21    -2    -9   -9
5   1     19    -9    -6   NA
6   1     20    -9   -12  -13

# performs MI, using imputation sequence: CHG2, CHG3
imp<-mice(hamdchng, seed=123,visitSequence=c(4,5))

# creates data with imputed sets stacked (if need to look
at data)
```

```
# columns are trt, basval CHG1, CHG2,CHG3
stacked<-complete(imp,"long")

# fits a linear model for changes in HAMD at vis3 to each
completed set
fit<-with(imp, lm(CHG3~trt+basval))

# pooled inference using Rubin's rules
est<-pool(fit)
summary(est)
```

Code Fragment 19.3   SAS Code for MI analysis. Combined inference using PROC MIANALYZE

```
/* combined inference for estimated treatment contrasts */
  ODS OUTPUT PARAMETERESTIMATES=ES_MI;
  PROC MIANALYZE DATA=ES_MI_M ;
   BY TIME;
   MODELEFFECTS ESTIMATE;
   STDERR STDERR;
  RUN;

/* combined inference for LSMEANS */
  ODS OUTPUT PARAMETERESTIMATES=LS_MI;
  PROC MIANALYZE DATA=LS_MI_M ;
   BY TIME TRT;
   MODELEFFECTS ESTIMATE;
   STDERR STDERR;
  RUN;
```

Code Fragment 19.4   SAS Code for MI analysis. Imputing data from non-monotone pattern using MCMC

```
/* using MCMC to impute data from non-monotone patterns */

PROC MI DATA = ALL2_TRANSP2 OUT = ALL2_MIOUT NIMPUTE=1000
SEED=123;
 BY TRT;
 MCMC CHAIN=SINGLE NBITER=200 NITER=100 INITIAL=EM
    (ITPRINT  MAXITER=200);
 VAR BASVAL YOBS1 YOBS2 YOBS3;
RUN;
```

```
/* using MCMC to first complete data to monotone and then
perform a single imputation to each monotone set by Bayesian
regression to produce complete data sets */

PROC MI DATA = ALL2_TRANSP2 OUT = ALL4_MIOUT NIMPUTE=1000
SEED=123;
   BY TRT;
   MCMC IMPUTE = MONOTONE;
   VAR BASVAL YOBS1 YOBS2 YOBS3;
RUN;

PROC SORT DATA=ALL4_MIOUT; BY  _IMPUTATION_ TRT; RUN;

PROC MI DATA = ALL4_MIOUT OUT = ALL5_
MIOUT NIMPUTE =1 SEED =123;
   BY _IMPUTATION_ TRT;
   MONOTONE METHOD=REG;
   VAR BASVAL YOBS1 YOBS2 YOBS3;
RUN;
```

Code Fragment 19.5 SAS Code for MI analysis. Imputing data for baseline covariates using MCMC

```
/* using MCMC to first impute baseline covariates X1, X2, X3,
basval and other missing values deviating from monotone
pattern and then perform a single imputation to each
monotoneset by Bayesian regression to produce complete
data sets */

PROC MI DATA = ALL2_TRANSP2 OUT = ALL4_MIOUT NIMPUTE=1000
SEED=123;
   BY TRT;
   MCMC IMPUTE = MONOTONE;
   VAR X1 X2 X3 BASVAL YOBS1 YOBS2 YOBS3;
RUN;

PROC SORT DATA=ALL4_MIOUT; BY _IMPUTATION_ TRT; RUN;

PROC MI DATA = ALL4_MIOUT OUT = ALL5_
MIOUT NIMPUTE=1 SEED=123;
   BY _IMPUTATION_ TRT;
   MONOTONE METHOD=REG;
   VAR X1 X2 X3 BASVAL YOBS1 YOBS2 YOBS3;
RUN;
```

Code Fragment 19.6   SAS Code for an inclusive MI strategy: joint imputation of changes in HAMD and PGIIMP

```
/*Imputing from joint multivariate normal distribution*/

  PROC MI DATA = ALL3_TRANSP  OUT = ALL3_MIOUT NIMPUTE=1000
  SEED=123;
  BY TRT;
  MONOTONE METHOD=REG;
  VAR BASVAL CHG1 PGIIMP1 CHG2 PGIIMP2 CHG3 ;
  RUN;


/*multiple imputation using normal regression for changes in
HAMD and ordinal logistic regression for PGIIMP*/

  PROC MI DATA = ALL3_TRANSP  OUT = ALL3_MIOUT NIMPUTE=1000
  SEED=123;
  BY TRT;
  CLASS PGIIMPGR1 PGIIMPGR2;
  VAR BASVAL CHG1 PGIIMPGR1 CHG2 PGIIMPGR2 CHG3;
  MONOTONE REG (CHG1=BASVAL);
  MONOTONE LOGISTIC (PGIIMPGR1 =CHG1 BASVAL);
  MONOTONE REG  (CHG2= PGIIMPGR1 CHG1 BASVAL);
  MONOTONE LOGISTIC (PGIIMPGR2= CHG2 PGIIMPGR1 CHG1 BASVAL);
  MONOTONE REG  (CHG3= PGIIMPGR2 CHG2 PGIIMPGR1 CHG1 BASVAL);
  RUN;
```

Code Fragment 19.7   Example of R Code for an inclusive MI strategy: joint imputation of changes in HAMD and PGIIMP

```
#### joint imputation of continuous and categorical (ordered) outcomes


hamdpgi<-widedata[,c("trt","basval","CHG1","PGIIMPGR1","CHG2",
"PGIIMPGR2","CHG3")]

# make PGIIMP ordered factors
hamdpgi$PGIIMPGR1<-ordered(hamdpgi$PGIIMPGR1)
hamdpgi$PGIIMPGR2<-ordered(hamdpgi$PGIIMPGR2)

require(mice)

imp<-mice(hamdpgi, seed=123, visitSequence=c(5,6,7))
```

```
# fits linear model for change in HAMD at Time 3 to each
imputed data set
fit<-with(imp, lm(CHG3~trt+basval))

# pooled inference using Rubin's rules
est<-pool(fit)
summary(est)
```

19.9 Summary

MI is a popular and accessible method of model-based imputation. The three basic steps to MI include imputation, analysis, and inference. These steps can be applied to different types of outcomes including continuous (e.g., imputed via multivariate normal distribution), categorical (e.g., imputed via logistic model), and time-to-event (e.g., imputed with piecewise exponential proportional hazard model) outcomes.

If MI is implemented using the same imputation and analysis model, and the model is the same as the analysis model used in an ML-based analysis, MI and ML will yield asymptotically similar point estimates, but ML will be somewhat more efficient. However, with distinct steps for imputation and analysis, MI has more flexibility than other methods. This flexibility can be exploited in several situations.

Scenarios where MI is particularly useful include those when covariates are missing, when data cannot be modeled via multivariate normal distribution such that likelihood-based analyses are difficult to implement, and when inclusive modeling strategies are used to help account for missing data. MI can be also very useful for sensitivity analyses.

20

Inverse Probability Weighted Generalized Estimated Equations

20.1 Introduction

Inverse probability weighting (IPW) has many applications. The focus here is on using IPW in conjunction with generalized estimating equations (GEEs) to construct weighted GEE (wGEE). A standard reference for wGEE is Robins, Rotnitzky, and Zhao (1995). For details, see also, for example, Molenberghs et al. (2015). The chapter begins with brief review of some technical details for GEE and IPW and then provides implementation examples.

20.2 Technical Details – Generalized Estimating Equations

Seminal references for GEEs are Liang and Zeger (1986) and Diggle, Heagerty, Liang, and Zeger (2002). Essentially, GEE allows for correlation between repeated measurements on the same subjects without explicitly defining a model for the origin of the dependency. Consequently, GEE is less sensitive to parametric assumptions than maximum likelihood and is computationally more efficient. Some connections can be drawn between the origins of GEE and maximum likelihood. However, GEE is not a likelihood-based method, and hence frequentist, and similar to least squares, the restrictive assumption of a missing completely at random (MCAR) missing data mechanism is generally required for ignorability of missing data in GEE.

In GEE, estimates arise as generalizations of both quasi-likelihood and generalized linear models from univariate to multivariate outcomes. It is a viable approach for conducting inference on parameters that can be expressed via the first moments of some underlying multivariate distributions (e.g., treatment contrast in repeated measures analysis), especially in situations when specification of the entire multivariate distribution may be challenging. For example, GEE can be useful for analysis of non-normally distributed (e.g., binary) correlated data when the maximum likelihood methods are usually hard to implement and/or require excessive computation time.

The attractiveness of GEE is in that it does not require modeling within-subject correlation structures. Even if the structure is incorrectly specified (e.g., assumed to be independent), the point estimates of parameters are consistent, and the correct standard errors can be computed using the robust sandwich estimator that is based on residuals from the fitted model. However, the relaxed distributional assumptions and nonreliance on correlation structure comes at the price of generally decreased statistical efficiency. That is, all else equal, parameter estimates from GEE will have greater standard errors than corresponding maximum likelihood estimates.

GEE obtains estimates as solutions of estimating equations that have the following form:

$$s(\beta) = \sum_{i=1}^{N} D_i^T V_i^{-1} \left(y_i - \mu_i(\beta) \right) = 0.$$

Here, the summation is over N subjects, y_i is a vector of observed n_i outcomes for the ith subject, μ_i is a vector of expected marginal means for the ith subject (i.e., the set of n_i visitwise marginal means μ_{ij} for the ith subject with evaluations at visits $j = 1,...,n_i$) that are expressed in terms of the linear combination $x'_{ij}\beta$ through an appropriate link function. For example, when modeling binary data using the logit link, the marginal means are probabilities of the event of interest that are related to a linear predictor $l_{ij} = x'_{ij}\beta$ via an inverse logit transformation $pr(Y_{ij} = 1) = \mu_{ij} = 1/(1+\exp(-l_{ij}))$. The linear combination l_{ij} is the product of row-vector x'_{ij} in the $N \times p$ design matrix X_j (for jth visit) and the p-dimensional vector of parameters β (e.g., treatment by visit interactions, baseline severity score).

The $n_i \times p$ matrix D_i contains partial derivatives μ_i with respect to parameters β and the $n_i \times n_i$ matrix V_i is essentially a "working model" for the covariance of Y_i that is decomposed into a so-called working correlation matrix $R(\alpha)$ (where α is the vector of estimated parameters) that is premultiplied and postmultiplied by the square root of diagonal matrix A_i with marginal variances.

$$V_i = \phi A_i^{\frac{1}{2}} R(\alpha) A_i^{\frac{1}{2}}$$

Note that the GEE contains p equations. In the simplest case of a univariate normal regression, these specialize to p "normal equations" that are obtained by differentiating the log-likelihood (or least squares) with respect to p parameters in β.

GEEs are not likelihood-based and parameter estimates are biased under MAR (Robins, Rotnitzky and Zhao, 1995). The stronger assumption of MCAR is required because the root of the estimating equation expectation is in general not equal to the true parameters underlying the outcome process, unless the missingness is MCAR. The requirement of MCAR stems from basing estimates of the parameters in the working correlation matrix on what is essentially a complete case analysis. Only subjects that were observed at both visits t_1 and t_2 contribute to the estimate of within-subject correlation between visits t_1 and t_2. When within-subject correlations are not used in computing point estimates (the case of "independent working correlations"), bias under MAR is even more apparent since GEE point estimates of treatment effect at each time point are merely treatment contrasts evaluated for complete cases at a given visit.

The bias in GEE resulting from MAR missingness can be removed by incorporating weights into the GEE (wGEE). The weights are based on the inverse probability (IP) of observing the dropout patterns that were present in the data. The idea of weighting traces back to the well-known Horvitz and Thompson (1952) IP estimator.

20.3 Technical Details – Inverse Probability Weighting

20.3.1 General Considerations

The motivation behind IPW is to correct for bias caused by nonrandom selection (dropout). Assume interest is in computing the expected value of a discrete random variable, Y, in a finite population of N elements, and Y can assume only two possible values: y_1 and y_2. The population mean is computed as a weighted average:

$$\mu = E(Y) = \frac{N_1 y_1 + N_2 y_2}{N}.$$

Now assume that a selection mechanism is applied to the population that removes some elements y_1 with probability of selection $\pi_1 = 1/2$ while not affecting the selection of y_2 ($\pi_2 = 1$). Therefore, only $\frac{N_1}{2}$ occurrences of y_1 are expected to be observed, while N_2 occurrences of y_2 are observed. Assume that the relative frequencies of y_1 and y_2 in the population equal these expected values (i.e., ignoring the randomness in the selection process). Then, when estimating the population mean based on the mean from incomplete data ("observed case" or "complete case" estimator, $\hat{\mu}_{cc}$), the expectation of this estimator is

$$E\left(\hat{\mu}_{cc}\right) = \frac{\frac{1}{2}N_1 y_1 + N_2 y_2}{\frac{1}{2}N_1 + N_2}.$$

This estimator is biased because in general $E(\hat{\mu}_{cc}) \neq E(Y)$. In this example, the selection probabilities are known and adjustments for the selection bias can be made by associating y_1 and y_2 with weights computed as the inverse of their respective selection probabilities $w_1 = \frac{1}{\pi_1} = 2, w_2 = \frac{1}{\pi_2} = 1$. Thus

$$E(\hat{\mu}_{ipw}) = \frac{\dfrac{1}{2}N_1 y_1 w_1 + N_2 y_2 w_2}{\dfrac{1}{2}N_1 w_1 + N_2 w_2} = \frac{\dfrac{1}{2}N_1 y_1 2 + N_2 y_2 1}{\dfrac{1}{2}N_1 2 + N_2 1} = \frac{N_1 y_1 + N_2 y_2}{N} = E(Y).$$

With $\pi_1 = 1/2$, the inverse is $2/1$, which essentially counts each observed value of y_1 twice because the probability of observing each instance of y_1 from the population is $1/2$. If $\pi_1 = 2/3$, then each occurrence of y_1 would be given a weight of $3/2 = 1.5$. In actual clinical trial data, the probabilities are unknown and must be estimated from sample data.

If the selection probability for elements y_1 were $\pi_1 = 0$, then all y_1 are missing, and IPW is not feasible. Hence, an important requirement of IPW is that all possible values of the outcome variable have nonzero probability of being in the observed population. If some values have very low probabilities, it is unlikely that these values would be observed in a sample. In such cases, IPW may lead to bias (if some values are not observed in the sample) and/or high variance (if some values are observed but have low estimated probabilities resulting in very large weights).

To illustrate a general use of IPW, let $y_i, i = 1,...,N$, denote realizations of an outcome variable Y in a complete data set, $y_{i,obs}, i = 1,...,N_1$ and r_i is the observed values of missingness indicator variable R so that for subject i.

$$y_i = \begin{cases} y_{i,obs}, \ if \ r_i = 1 \\ y_{i,mis}, \ if \ r_i = 0 \end{cases}$$

Assume that the true selection probabilities (or probabilities of nonmissingness) are known, $\pi_i = \Pr(R_i = 1)$. Note that in general π_i may be a function of y_i, therefore π_is are conditional probabilities $\pi_i = \Pr(R_i = 1 | Y = y_i)$. Define the IPW estimator as

$$\hat{\mu}_{ipw} = \frac{1}{N}\sum_{i=1}^{N_1} y_{i,obs}\frac{1}{\pi_i}.$$

Equivalently, in terms of the complete data outcome as

$$\hat{\mu}_{ipw} = \frac{1}{N}\sum_{i=1}^{N} y_i r_i \frac{1}{\pi_i}.$$

Although not all y_i are observed, the product $y_i r_i$ is fully observed. To show that $\hat{\mu}_{ipw}$ is an unbiased estimator of $\mu = E(Y)$, its expectation with respect to the joint distribution of Y and R is needed. The law of conditional iterated expectations can be exploited to allow decomposition of the joint expectation as $E_{Y,R}(\cdot) = E_Y\{E_{R|Y}(\cdot)\}$. Then

$$E\left(\hat{\mu}_{ipw}\right) = \frac{1}{N}\sum_{i=1}^{N}E(Y_iR_i)\frac{1}{\pi_i} = \frac{1}{N}\sum_{i=1}^{N}\frac{1}{\pi_i}E(Y_iE_{R|Y}R_i)$$

$$= \frac{1}{N}\sum_{i=1}^{N}\frac{1}{\pi_i}\Pr\left(R=1|\, Y=y_i\right)E(Y_i) = \frac{1}{N}\sum_{i=1}^{N}\frac{\pi_i}{\pi_i}E(Y_i)$$

$$= \frac{1}{N}\sum_{i=1}^{N}E(Y_i) = E(Y) = \mu$$

Therefore, assigning weights that are the IP of observing the outcomes recovers the mean expected in the complete data. In practice, the weighted sum of observed values is divided by the sum of weights rather than by N.

$$\hat{\mu}_{ipw} = \frac{1}{\sum_{i=1}^{N_1}\frac{1}{\pi_i}}\sum_{i=1}^{N_1}y_{i,obs}\frac{1}{\pi_i}$$

In expectation, these are equivalent: $E\left(\sum_{i=1}^{N_1}\frac{1}{\pi_i}\right) = E\left(\sum_{i=1}^{N}\frac{R_i}{\pi_i}\right) = E\left(\sum_{i=1}^{N}\frac{\pi_i}{\pi_i}\right) = N.$ Therefore, the relative subject-specific weights (summing up to 1) can be defined as

$$\hat{\mu}_{ipw} = \sum_{i=1}^{N_1}y_{i,obs}w_i$$

$$w_i = \frac{\frac{1}{\pi_i}}{\sum_{i=1}^{N_1}\frac{1}{\pi_i}}$$

Once IPW estimates of the expected outcome for each treatment group $\left(\hat{\mu}_{ipw}^{(t)}, t = 0,1\right)$ are obtained, the treatment effects can be estimated as $\hat{\theta}_{ipw} = \hat{\mu}_{ipw}^{(1)} - \hat{\mu}_{ipw}^{(0)}.$

As an initial illustration, consider estimating treatment effects via inverse probably weighting of the completers (subjects with no missing data) in the small example data set with dropout. For this illustration, subjects are considered to have only one observation, the Time 3 assessment. The IPW is not known but can be estimated using logistic regression. For this example, the probability of dropout was estimated using a model applied to data pooled across visits with the dropout indicator as the response variable. Predictor variables included change from baseline at the previous visit, baseline score, and treatment. The inverse weights estimated from this model were applied to the subset of subjects that completed the trial (sometimes termed IP weighted complete case or IPWCC estimator) in order to estimate the mean that would have been observed if all subjects had completed.

For this example, the probabilities of completing the trial are computed as a product of conditional probabilities of completing each of the visits. That is,

$$\pi_i = Pr\left(R_{1i} = 1\right)Pr(R_{2i} = 1 \mid R_{1i} = 1)Pr(R_{3i} = 1 \mid R_{1i} = 1, R_{2i} = 1).$$

Because all subjects completed the first post-baseline visit, it is assumed the $Pr\left(R_{1i} = 1\right) = 1$. Estimates of conditional probabilities are obtained as functions of observed outcomes and other baseline covariates (e.g., using logistic regression) as will be shown in the next sections.

Weighting removes selection bias by giving larger weight to outcomes that are underrepresented in the observed data compared to what would have been observed if there were no missing data. Conversely, smaller weights are given to observed values that are overrepresented. Typically, as is the case with the small example data set with dropout, subjects with poorer efficacy are less likely to be observed. Therefore, subjects with poor outcomes are underrepresented in the observed data, and their weight should be larger compared to those who had larger improvements.

As expected, weights estimated from the small example data set are larger for subjects with worse outcomes (changes from baseline closer to zero) and smaller for subjects with better outcome (larger negative changes from baseline). The left-hand panel of Figure 20.1 illustrates this relationship. The right panel is a plot of visitwise mean changes from baseline for subjects who competed the trial divided into two strata; the first stratum is subjects whose estimated weights were less than the median and the second stratum is subjects with weights larger than the median. The figure clearly shows subjects with larger weights had worse marginal means than those with smaller weights.

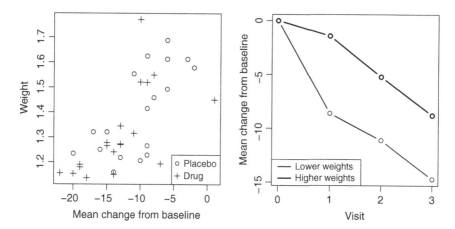

FIGURE 20.1

Relationship between weights and changes from baseline for completers. Left panel is scatter plot of subject weights versus changes from baseline to visit 3. Right panel is mean change profiles for completers with weights larger than median (thick line) and lower than the median (thin line).

20.3.2 Specific Implementations

The previous section illustrated IPW for an analysis of covariance (ANCOVA) model with a single outcome per subject. Here, IPW is considered in the context of repeated measures analysis via wGEE. IP weights are commonly calculated in one of the two ways. Either one weight per subject that is applied to all visits, thus reflecting the probability of the observed dropout pattern, or each subject gets a unique weight at each visit that reflects the visitwise probabilities of dropout.

Robins and Rotnitzky (1995) proposed IP weights that change by visit (observation-level weighting). The weights are incorporated into the working covariance model as the elements of a diagonal $n_i \times n_i$ weight matrix W_i

$$V_i = \phi A_i^{\frac{1}{2}} W_i^{-\frac{1}{2}} R(\alpha) W_i^{-\frac{1}{2}} A_i^{\frac{1}{2}}.$$

The weights are computed for each subject at each time point in a manner similar to how the weights for the last visit were computed in the weighted analysis of completers in the previous section. For example, Subject 9 had two completed visits; hence, the matrix W is a 2×2 diagonal matrix with weights based on estimated probabilities as follows (subject index omitted whenever it does not cause confusion):

$$w_1 = \frac{1}{\hat{\pi}_1}, w_2 = \frac{1}{\hat{\pi}_2}$$

$$\hat{\pi}_1 = \widehat{Pr}(R_1 = 1 \mid x)$$

$$\hat{\pi}_2 = \widehat{Pr}(R_1 = 1 \mid X)\widehat{Pr}(R_2 = 1 \mid R_1 = 1, x, y_1)$$

Here $\hat{\pi}_1 = 1$, and correspondingly, $\hat{\pi}_2 = \widehat{Pr}(R_2 = 1 \mid R_1 = 1, x, y_1)$, because no dropouts had occurred prior to the first post-baseline visit. (Note we use $Pr(R_2 = 1 \mid R_1 = 1, x, y_1)$ as a shorthand for $Pr(R_2 = 1 \mid R_1 = 1, X = x, Y_1 = y_1)$). The weights can change from visit to visit reflecting the changing probability of a subject remaining in the study given the previous data.

Fitzmaurice, Molenberghs, and Lipsitz (1995) proposed subject-level weighting with a single weight for each subject that is used at every visit (repeated for all diagonal elements of the W_i matrix). This weight is an inverse of the estimated probability $\hat{\pi}^{(d)}$ of the dropout pattern d observed for that subject. Here the pattern $d = 1, \ldots, T+1$ indicates the *next* time point after the subject was observed for the last time. That is, for a subject who was last seen at time t, $d = t+1$. For completers, $d = T+1$, where T indicates the last scheduled visit.

The working covariance model is

$$V_i = \hat{\pi}_i^{(d)} \phi A_i^{\frac{1}{2}} R(\alpha) A_i^{\frac{1}{2}}.$$

We obtain the weighted estimating equations from the original estimating equations by multiplying each subject's contributions to the estimating equations by the IP of his or her observed dropout pattern $w_i = 1 / \hat{\pi}_i^{(d)}$.

To illustrate, for a subject that completed all the scheduled visits, the weight is computed exactly as weights for completers at the last visit were computed in the previous section. That is,

$$\hat{\pi}^{(d)} = \widehat{Pr}(R_1 = 1 \mid x)\widehat{Pr}(R_2 = 1 \mid R_1 = 1, x, y_1)\widehat{Pr}(R_3 = 1 \mid R_1 = 1, R_2 = 1, x, y_1, y_2).$$

For subjects who did not complete the trial, the probability of the observed pattern is computed as the probability of that subject remaining in the trial until the last observed visit and discontinuing right after. As an example, for Subject 9, the probability of the observed dropout pattern is

$$\hat{\pi}^{(d)} = \widehat{Pr}(R_1 = 1 \mid x)\widehat{Pr}(R_2 = 1 \mid R_1 = 1, x, y_1)\left[1 - \widehat{Pr}(R_3 = 1 \mid R_1 = 1, R_2 = 1, x, y_1, y_2)\right].$$

20.4 Example

With the building blocks of IPW and GEE in place, a wGEE analysis of the small example data set with dropout can be considered. The goal is to estimate treatment effects using a wGEE model, with weights reflecting the selection probability, that is, the probability of completing all three visits. This example uses the approach of Fitzmaurice, Molenberghs, and Lipsitz (1995) where subject weights are computed based on the IP of observed dropout pattern $\hat{\pi}_i^{(d)}$.

After estimating the probabilities $\hat{\pi}_i$, the weights are computed as $w_i = \frac{1}{\hat{\pi}_i^{(d)}}$. The weights are then included in the data set as additional variable, and the weighted analysis is conducted, for example, via SAS PROC GENMOD (SAS, 2013). The weights are data dependent, but the analysis assumes the weights are known, similar to the assumption that variance components are known in a standard mixed-effects model analysis. The robust sandwich estimates of standard errors provided by the GEE are still valid although conservative (Robins, Rotnitzky, and Zhao, 1995). Resampling methods can be used to obtain nonconservative estimates.

It is important to recognize that wGEE provides valid estimates under MAR – assuming correct specification of the dropout model. As previously noted, estimating weights can be challenging when some of the true probabilities of observed patterns are close to 0. This can result in large weights and unstable estimates of treatment effects. Truncation and stratification can be used to stabilize weights. Moreover, weightings that are more stable have been proposed (Hernán, Brumback, and Robins, 2000).

SAS code for obtaining subject-level and observational-level IP weights is listed in Code Fragment 20.1 (Section 20.5). SAS code to implement wGEE analyses via PROC GENMOD using the calculated IPW weights is listed in Code Fragment 20.2 (Section 20.5), followed by Code Fragment 20.3 illustrating wGEE analysis via the experimental procedure PROC GEE that is available in SAS/STAT 13.2 (SAS, 2013) wherein the weights are obtained as part of the PROC.

Results for the wGEE analyses using subject-level weights are compared to results from an unweighted GEE and results from complete data in Table 20.1. It is important to interpret these results as just one realization from a stochastic process. Replicating the comparisons many times under the same conditions – including missing data arising from an MAR mechanism – the average of the LSMEANS and treatment contrasts across the repeated samples would be asymptotically equal for complete data, incomplete data analyzed via direct likelihood, MI, and wGEE with an appropriate dropout model. Standard errors would be consistently greater in incomplete data.

The wGEE analysis yielded a point estimate for the treatment contrast at Time = 3 between the result obtained from the analysis of complete cases and from complete data. As expected, the standard error for the Time 3 treatment

TABLE 20.1

Results from GEE and wGEE Analyses of the Small Example Data Set

Treatment	Time	Complete Data		GEE (Incomplete Data)		wGEE (Incomplete Data)	
		LSMEANS	SE	LSMEANS	SE	LSMEANS	SE
1	1	−4.13	0.91	−4.10	0.76	−4.38	0.83
1	2	−6.70	0.93	−6.70	1.04	−5.64	0.87
1	3	−9.86	1.05	−10.17	1.11	−9.58	1.08
2	1	−5.32	0.91	−5.29	0.97	−4.88	0.88
2	2	−8.70	0.93	−8.29	0.96	−6.09	1.10
2	3	−13.26	1.05	−13.10	1.27	−12.71	1.28
Endpoint Treatment Difference		3.39	1.49 ($p = 0.0274$)	2.92	1.73 ($p = 0.0905$)	3.13	1.71 ($p = 0.0670$)

contrast from wGEE with incomplete data was larger than the standard error from complete data and slightly larger than the standard error from the direct likelihood analysis of incomplete data (see Table 18.1). The relaxed distributional assumptions of GEE result in a slightly less efficient analyses than likelihood-based methods.

20.5 Code Fragments

Code Fragment 20.1   SAS code for obtaining IP weights

```
/* Create change at previous visit variable and dropout
indicator variable */
PROC SORT DATA=ALL2; BY SUBJECT TIME; RUN;
DATA FOR_WGEE;
 SET ALL2;
 RETAIN DROP CHANGE_LAG1;
 BY SUBJECT TIME;
 IF FIRST.SUBJECT THEN DO;
  DROP=0;
     CHANGE_LAG1=.;
 END;
 IF DROP=1 THEN DELETE;
 IF CHGDROP=.THEN DROP=1;
 OUTPUT;
 CHANGE_LAG1=CHGDROP;
RUN;

/* Logistic regression analysis to obtain probabilities */
PROC LOGISTIC DATA=FOR_WGEE DESC;
 CLASS TRT;
 MODEL DROP=TRT BASVAL CHANGE_LAG1;
 OUTPUT OUT =PRED P=PRED;
RUN;

/* Calculate and merge weights into data set */
 DATA WEIGHTS_SUB (DROP=WEIGHT_OBS PB_OBS )
  WEIGHTS_OBS (DROP=WEIGHT_SUB PB_SUB);
SET  PRED;
BY SUBJECT TIME;
RETAIN PB_SUB PB_OBS;
IF FIRST.SUBJECT THEN DO;
 PB_OBS =1;  /* PROB. OF OBSERVING SUBJECT AT GIVEN TIME -
  FOR OBSERVATIONAL WEIGHTS */
 PB_SUB =1;  /* PROB. OF OBSERVED DROPOUT PATTERN -
  FOR SUBJECT WEIGHTS */
END; ELSE DO;
```

```
IF DROP THEN PB_SUB =PB_SUB * PRED; ELSE PB_SUB =
 PB_SUB *(1-PRED);
 PB_OBS =PB_OBS * (1-PRED);
END;
WEIGHT_OBS=1/PB_OBS ;
OUTPUT WEIGHTS_OBS;
IF LAST.SUBJECT THEN DO;
 WEIGHT_SUB =1/PB_SUB;
 OUTPUT WEIGHTS_SUB;
END;
RUN;


PROC SORT DATA=FOR_WGEE; BY SUBJECT; RUN;
PROC SORT DATA=WEIGHTS_SUB; BY SUBJECT; RUN;


DATA FOR_WGEE;
MERGE FOR_WGEE
  WEIGHTS_SUB (KEEP=SUBJECT WEIGHT_SUB);
BY SUBJECT;
RUN;
```

Code Fragment 20.2   SAS code for wGEE analysis using the PROC GENMOD

```
ODS OUTPUT ESTIMATES=ES_CHDR_WGEE_IND
      LSMEANS=LS_CHDR_WGEE_IND;
PROC GENMOD DATA=FOR_WGEE (WHERE=(CHGDROP NE.));
 CLASS TRT TIME SUBJECT;
 MODEL CHGDROP= BASVAL TRT TIME TRT*TIME /LINK=ID DIST=NORMAL;
 WEIGHT WEIGHT_XXX¹;
 REPEATED SUBJECT=SUBJECT /TYPE=IND;
 ESTIMATE "TRT EFFECT AT VIS 3" TRT -1 1 TRT*TIME 0 0
  -1 0 1;
 LSMEANS TRT*TIME;
RUN;
```

  1. Use WEIGHT _ SUB for subject-level weights and WEIGHT _ OBS
     for observation-level weights

Code Fragment 20.3   SAS code for wGEE analysis using the experimental PROC GEE

```
PROC GEE DATA=ALL2 DESC PLOTS=HISTOGRAM;
 CLASS SUBJECT TRT TIME;
 MISSMODEL PREVY TRT BASVAL / TYPE=SUBLEVEL;
 MODEL CHGDROP = BASVAL TRT TIME TRT*TIME BASVAL*TIME /
DIST=NORMAL LINK=ID;
 REPEATED SUBJECT=SUBJECT / WITHIN=TIME CORR=IND;
RUN;
```

20.6 Summary

Although GEE is valid only under MCAR, IPW can correct for MAR, provided an appropriate model for the missingness process (dropout) is used whereby missingness depends on observed outcomes but not further on unobserved outcomes. The weights are based on the IP of dropout and in effect create a pseudo-population of data that would have been observed with no missing values in an infinitely large trial.

Weighting can be at the subject level, with one weight per subject reflecting the IP of observing the dropout pattern; or weighting can be at the observation level, with one weight per subject per visit that reflects the changing probability of dropout as outcomes evolve over time.

As with standard GEE, in wGEE, no assumptions about the correlation structure are required and therefore wGEE yields semiparametric estimators. The wGEE estimates are generally not as efficient as maximum likelihood (parametric) estimators obtained using the correct model, but they remain consistent, whereas the maximum likelihood estimators from a misspecified parametric model are inconsistent.

21

Doubly Robust Methods

21.1 Introduction

Common references for the emerging area of doubly robust methods include Carpenter, Kenward, and Vansteelandt (2006) and Tsiatis (2006). Doubly robust methods can be viewed as an extension of generalized estimating equation (GEE). Recall (Chapter 20) that GEE is valid only under missing completely at random (MCAR), and inverse probability weighting (IPW) can correct for MAR, provided an appropriate model for the missingness process (dropout) is used whereby missingness depends on observed outcomes but not further on unobserved outcomes (Molenberghs and Kenward, 2007). The method of weighted GEE (wGEE) yields semiparametric estimators because it does not model the entire distribution of the outcome variable. These semiparametric estimates are generally not as efficient as maximum likelihood estimators obtained using the correct model, but they remain consistent when the maximum likelihood estimators from a misspecified parametric model are inconsistent (Molenberghs and Kenward, 2007). Therefore, a motivation for doubly robust methods is to improve the efficiency of wGEE by augmenting the estimating equations with the predicted distribution of the unobserved data given the observed data (Molenberghs and Kenward, 2007).

In addition to improving efficiency, augmentation introduces the property of double robustness. To understand double robustness, first consider that efficient IPW estimators require three models. (1) The substantive (analysis) model relates the outcome to explanatory variables and/or covariates. (2) A model for the probability of observing the data (usually a logistic model of some form). (3) A model for the joint distribution of the partially and fully observed data, which is compatible with the substantive model in (1) (Molenberghs and Kenward, 2007).

If model (1) is wrong, for example, because a key confounder is omitted, estimates of all parameters will typically be inconsistent. The intriguing property of augmented wGEE is that if either model (2) or model (3) is wrong, but not both, the estimators in model (1) are still consistent (Molenberghs and Kenward, 2007). Although doubly robust methods have received considerable attention, they haven't been studied and applied as much as other methods. Readers can refer to Robins, Rotnitzky, and Zhao (1995), Carpenter, Kenward, and Vansteelandt (2006), Tsiatis (2006), and Daniel and Kenward (2012) for additional details.

21.2 Technical Details

In Section 20.2, inverse probability estimation was explained starting with the simple case of estimating the sample mean of an outcome variable Y when some of its values y_i, $i = 1,...,N$ are missing (indicated with $r_i = 0$). Subsequently, IPW was extended to repeated measures. A similar approach is used here, with the ideas of doubly robust estimation building on the wGEE results of Chapter 20.

To fix ideas, assume that missingness is MAR and the probability that Y is missing does not depend on unobserved vales of Y, given covariates $\mathbf{X}$ (which includes various pretreatment covariates). Recall that an unbiased estimate of treatment means can be obtained via the inverse probability estimator:

$$\hat{\mu}_{ipw} = \frac{1}{N} \sum_{i=1}^{N} y_i r_i \frac{1}{\pi_i},$$

where π_i are probabilities of having observed outcome (known or estimated from available data). Under MAR, $\pi_i = \pi(x_i) = \Pr(R = 1 \mid X = x_i)$ is a known or estimated function of covariates; therefore, $\pi = \pi(x)$. Consistency of the IPW estimator follows from the assumption that the model for probability of missingness was correctly specified. If that is not the case, probabilities $\hat{\pi}_i$ estimated from an incorrect model will converge (as N becomes large) to some "misspecified" function $\ddot{\pi}(x)$ and the estimate of the population mean would (in general) be biased. The *augmented inverse probability weighted estimator* of sample mean $\hat{\mu}_{aipw}$ aims to protect against bias due to misspecification of the weights by combining IPW with an imputation model, $m(x)$.

$$\hat{\mu}_{aipw} = \frac{1}{N} \sum_{i=1}^{N} \tilde{y}_i = \frac{1}{N} \sum_{i=1}^{N} \left\{ \frac{r_i}{\hat{\pi}_i} y_i - \left(\frac{r_i}{\hat{\pi}_i} - 1 \right) \hat{m}_i \right\}$$

This requires estimation of models for (1) probability of being observed and (2) outcome value (whether missing or observed) by placing hats over $\hat{\pi}_i$ and $\hat{m}_i$. Therefore, the estimator is an average of N "augmented values," $\tilde{y}_i$, $i = 1,..., N$. Unlike the IPW estimator, the augmented IPW (AIPW) estimator receives nonzero contribution from subjects with missing values ($r = 0$), specifically such subjects contribute with their imputed or predicted value, $\widehat{m}_i$, estimated from the outcome model $m(x)$.

The AIPW estimator is written in the following alternative forms that are re-arrangements of the original estimator:

$$\widehat{\mu}_{aipw} = \frac{1}{N}\sum_{i=1}^{N}\left\{\frac{r_i}{\widehat{\pi}_i}y_i - \left(\frac{r_i}{\widehat{\pi}_i}-1\right)\widehat{m}_i\right\} = \frac{1}{N}\sum_{i=1}^{N}\left\{\frac{r_i}{\widehat{\pi}_i}\left(y_i - \widehat{m}_i\right) + \widehat{m}_i\right\}$$

$$= \frac{1}{N}\sum_{i=1}^{N}\left\{y_i + \left(\frac{r_i}{\widehat{\pi}_i}-1\right)\left(y_i - \widehat{m}_i\right)\right\}.$$

The last representation is particularly useful for demonstrating the *double robustness* property of the AIPW estimator: the estimator is consistent if at least one of the two models is correctly specified. To illustrate, assume that the sample size is large so that the estimators $\widehat{\pi}_i$ and $\widehat{m}_i$ converge (in probability) to their large sample counterparts, whether they are correctly or incorrectly specified functions of observed data. Also assume that the model for the probability of missingness is correctly specified, $\pi(x)$, and the outcome model is a (possibly) misspecified function $\ddot{m}(x)$. Applying the law of conditional iterated expectations to a joint distribution of random variables R, X, Y as $E_{R,X,Y}(\cdot) = E_{X,Y}\{E_{R|X,Y}(\cdot)\}$, the expectation of individual augmented values can be written as follows (after factoring out $Y - \ddot{m}(X)$ that is a constant conditional on X and Y):

$$E\left(\tilde{Y}\right) = E(Y) + E\left\{E\left(\frac{R}{\pi(X)}-1 \mid X, Y\right)(Y - \ddot{m}(X))\right\}$$

$$= \mu + E\left\{\left(\frac{E(R|X,Y)}{\pi(X)}-1\right)(Y - \ddot{m}(X))\right\}$$

$$= \mu + E\left\{\left(\frac{\pi(X)}{\pi(X)}-1\right)(Y - \ddot{m}(X))\right\}$$

$$= \mu + 0$$

The second term vanishes regardless of $E_{XY}(Y - \ddot{m}(X))$ because under MAR $E(R|X=x, Y=y) = Pr(R=1|X=x) = \pi(x)$, resulting in consistent estimation of the population mean μ.

Similarly, assume that the model for the probability of missingness is (possibly) misspecified, $\ddot{\pi}(x)$, and the outcome model is a correctly specified function $m(x)$, after applying the law of conditional iterated expectations, $E_{R,X,Y}(\cdot) = E_{R,X}\{E_{Y|R,X}(\cdot)\}$, the expectation of individual augmented values can be written as follows:

$$E(\tilde{Y}) = E(Y) + E\left\{\left(\frac{R}{\pi(X)} - 1\right)E(Y - m(X)|R,X)\right\}$$

$$= \mu + E\left\{\left(\frac{R}{\pi(X)} - 1\right)(E(Y|X,R) - m(X))\right\}$$

$$= \mu + E\left\{\left(\frac{R}{\pi(X)} - 1\right)(m(X) - m(X))\right\}$$

$$= \mu + 0.$$

Again, the second term vanishes in expectation because under MAR, the distribution of outcomes conditional on observed data is the same for subjects with observed ($R = 1$) and unobserved ($R = 0$), hence $E(Y|X = x, R = r) = E(Y|X = x) = m(x)$.

To summarize, the idea of doubly robust estimation is to combine the strengths of two models: for probability of missingness and for imputing missing values. Both models are estimated under the assumption of MAR utilizing observed subject-level data, which may be undesirable or awkward to include in the direct maximum likelihood-based modeling of the parameter of interest (e.g., treatment contrast at specific time point). The doubly robust estimation via AIPW results in (1) protection against one of the two models being misspecified (doubly robustness) and (2) increased efficiency. While the increase in efficiency may appear a natural outcome of using additional information (predicted values for subjects with missing outcomes), it relies on sophisticated theory of semiparametric estimation (Robins and Rotnitzky, 1995; Tsiatis, 2006), which is outside the scope of this book.

Despite the intuitive appeal, doubly robust estimators have received some criticism. For example, in situations when both models may be "slightly" misspecified (a typical situation with real data), the doubly robustness property may not translate to any gain, especially given inherent instability of IPW. For example, as argued in Kang and Schafer (2009), "two wrong models are not better than one" (see also a different view in Cao et al. (2009) and Tsiatis et al. (2011), who proposed improved double robust estimators).

As with IPW, AIPW estimator of a population mean gives rise to AIPW-based (generalized) estimating equations. Their general form starts with IPW estimating equations and adds some function .of the data with zero expectation:

$$\sum_{i=1}^{N}\left\{\frac{r_i}{\pi_i}U\left(y_i,x_i,|\,\theta\right)-\left(\frac{r_i}{\pi_i}-1\right)\phi\left(y_{i.obs},x_i\right)\right\}=0.$$

Like the AIPW estimator of the population mean, the AIPW estimating equations have two terms: one is based on contributions from observed data only (inversely weighted by π_i) and the second is based on contributions from both observed and unobserved cases. The zero mean function $\phi(y,x)$ is constructed to have minimal variance following the semiparametric theory of Robins and Rotnitzky (1995), resulting in conditional expectation of the score function $U(y_i,x_i|\,\theta)$ on the observed data Y_{obs}, X, which can be loosely termed "imputation model." However, implementations, especially for longitudinal data, are not straightforward and selection of the best augmented function may be not obvious. Carpenter, Kenward, and Vansteelandt (2006) provide an example of doubly robust estimators in the context of missing covariates, and Tsiatis et al. (2011) illustrate modeling of longitudinal outcomes.

21.3 Specific Implementations

Specific implementation of AIPW requires selection of (a) a substantive model of interest, (b) a missing data model for estimating IPW, and (c) an outcome (imputation) model. In the context of data analysis from clinical trials, the substantive model would typically evaluate the treatment effect (e.g., treatment contrast at the last scheduled visit). The logistic regression can be used to model the missingness process. A repeated measures analysis would typically be used to estimate the outcome model. Both (b) and (c) can be modeled using "inclusive" strategies, incorporating many covariates, whereas model (a) typically focuses on the primary estimand (parameter) of interest.

Consider a simple application of AIPW estimating equations in the context of the small example data set that has three post-baseline visits, which is a natural extension of the example of the IPW estimator of treatment effect in completers (IPWCC) from Section 20.3.1 (see Seaman and Copas, 2009 for other implementations).

Here the substantive model (a) is an analysis of covariance (ANCOVA) for Time 3 outcomes with terms for intercept, baseline severity covariate (Y_0), and treatment indicator (Z):

$$E(Y_3 \mid Z, Y_0) = \beta_0 + \beta_1 Z + \beta_2 Y_0.$$

Let x_i be a row vector comprising the constant 1 and two covariates $x_i = \{1, y_{0i}, z_i\}$. A possible augmented estimating equation can be

$$\sum_{i=1}^{N} \left\{ \frac{r_i}{\hat{\pi}_{i3}} x_i^T \left(y_{i3} - x_i^T \beta \right) - \left(\frac{r_i}{\hat{\pi}_{i3}} - 1 \right) \hat{E} \left(X_i^T \left(Y_{i3} - X_i^T \beta \right) \mid x_i, y_{i1}, y_{i2} \right) \right\} = 0.$$

Here (similarly to computing IPW weights for completers for the example in Section 20.3.1), the probability of being observed at the last visit $\hat{\pi}_{i3}$ is estimated as a function of baseline and available post-baseline data:

$$\pi_3 \left(x_i, y_{i.obs} \right) = \Pr(R_{i3} = 1 \mid R_{i2} = 1, x_i, y_{i1}, y_{i2}) \Pr(R_{i2} = 1 \mid x_i, y_{i1}).$$

The probabilities in the chain product can be estimated from separate logistic regressions for dropouts at Times 2 and 3 (recall that there are no dropouts the first post-baseline visit) $\Pr(R_{i1} = 1) = 1$.

Solving the above augmented estimating equations can be challenging. Let $\hat{m}_{i3} = \hat{E}(Y_{i3} \mid x_i, y_{i1}, y_{i2})$. The equations can be written by rearranging terms as

$$\sum_{i=1}^{N} \left\{ \frac{r_{i3}}{\hat{\pi}_{i3}} x_i^T \left(y_{i3} - \hat{m}_{i3} \right) + x_i^T \left(\hat{m}_{i3} - x_i^T \beta \right) \right\} = 0.$$

If the outcome model $m_3(x, y_1, y_2)$ is correctly specified, the first term vanishes in expectation and the roots of the equations (in expectation) must be the true values of the parameter vector β. With some choices of the estimated mean function $\hat{m}_3$, the elements of the first term (shown below as a 3×1 vector S) can be made *exactly* equal to zero, which, as seen shortly, would greatly simply estimation of parameters β:

$$S = \sum_{i=1}^{N} \left\{ \frac{r_{i3}}{\hat{\pi}_{i3}} x_i^T \left(y_{i3} - \hat{m}_{i3} \right) \right\} = 0.$$

Obviously, $S = 0$ can be enforced by choosing $\hat{m}_{i3}$ to be the solutions for the IPW estimating equations above. However, $\hat{m}_{i3}$ is sought as a function of a broader set of covariates *including all observed post-baseline outcomes,* not just those included in the covariate set X. To this end, m_{i3} is estimated as $\widehat{m}_{i3}^{\,ipw} = \widehat{m}_3(x_i, y_{i1}, y_{i2})$ by the weighted estimating equations as a linear function in X, Y_1, Y_2:

$$\sum_{i=1}^N \left\{ \frac{r_{i3}}{\widehat{\pi}_{i3}} (x_i, y_{i1}, y_{i2})^T \left(y_{i3} - m_3(x_i, y_{i1}, y_{i2}) \right) \right\} = 0.$$

Now that the elements of S are made exactly zero, the original augmented equations for estimating β's simplify to (unweighted) estimating equations:

$$\sum_{i=1}^N x_i^T \left(\widehat{m}_{i3}^{\,ipw} - x_i^T \beta \right) = 0,$$

where $\widehat{m}_{i3}^{\,ipw}$ are predicted outcomes at post-baseline visit 3 for each subject, either with observed or missing outcomes.

However, one caveat is that the predictive model m_3 includes as covariates (possibly missing) intermediate outcomes at Time 2, Y_2. Now we can apply (recursively) the same idea and use similar weighted estimating equations for estimating $\widehat{m}_{i2}^{\,ipw}$.

$$\sum_{i=1}^N \left\{ \frac{r_{i2}}{\widehat{\pi}_{i2}} (x_i, y_{i1})^T \left(y_{i2} - m_2(x_i, y_{i1}) \right) \right\} = 0.$$

Here, $\widehat{\pi}_{i2}$ are estimated probabilities of ith subject not missing at Time 2. Since no outcomes Y_1 are missing at Time 1, our "recursion" stops here and we will be able to compute $\widehat{m}_{i3}^{\,ipw} = \widehat{m}_3(x_i, y_{i1}, y_{i2}), i = 1, ..., N$, where the missing values for Y_2 will be replaced with predicted $\widehat{m}_{i2}^{\,ipw} = \widehat{m}_2(x_i, y_{i1})$.

This gives rise to the following general algorithm, suggested by Vansteelandt et al. (2010) for analysis of data with missing covariates and similar to that implemented for repeated measures recently in O'Kelly and Ratitch (2014, Section 8.4, pp. 377–378).

1. Compute subject-specific probabilities of being observed at the specific time point $\widehat{\pi}_{it}(x_1, x_2)$ using a full set of covariates partitioned into two subsets: X_1 (intended for modeling treatment effect) and X_2 (including baseline and post-baseline variables observed prior to the time point t when treatment assessment is made).

2. Predict outcome $\widehat{m}_{it}^{\,ipw}(x_1, x_2)$ at time point t for all subjects, whether observed or missing the outcome, using the full set of covariates and inverse probability weights from step 1 (e.g., via an appropriate

generalized linear model (GLM)). If some post-baseline covariates in the set X_2 are missing, iterate Steps 1 and 2 to impute them taking advantage of monotone patterns of missing data.

3. Use predicted values $\widehat{m}_{it}^{ipw}$ from Step 2 as new responses to fit a GLM model on covariates from the set X_1 and obtain estimates of the coefficients.

4. Compute standard errors for coefficients of the model in Step 3 via bootstrapping the entire modeling strategy (steps 1–3).

Details of this analytic strategy are illustrated in the next section via application to the example data set.

21.4 Example

In this section, doubly robust estimation as outlined in Section 21.3 is applied to the small example data set to evaluate the treatment effect at the last scheduled visit. This analysis involves the following steps (the SAS code is included in Code Fragment 21.1 in Section 21.5):

1. Estimating probabilities of discontinuation (and remaining on treatment) at Time 2 and 3 via logistic regression models. Specifically, the model for probability $\pi_{i2} = \widehat{Pr}(R_{i2} = 1 \mid x_i, y_{i1})$ of remaining on treatment by Time 2 included terms for baseline severity score, changes in outcome from baseline at Time 1, treatment indicator, and all interactions of treatment with other covariates. Only subjects who completed the first post-baseline visit contributed to this estimate (which in our case were all randomized subjects). Probability of remaining on treatment at Time 3 was estimated as a product of conditional probabilities $\widehat{\pi}_{i3} = \widehat{\pi}_{i2} \times \widehat{Pr}(R_{i3} = 1 \mid R_{i2} = 1, x_i, y_{i1}, y_{i2})$, where the second conditional probability was estimated from a separate logistic regression with terms for the baseline score, changes from baseline at Time 1 and 2, treatment indicator, and all treatment-by-covariate interactions. Only subjects completing the first two post-baseline visits contributed to this model.

2. Imputing missing values for the changes in outcome variable at Time 2 with its predicted values from the linear regression model $m_2(x_i, y_{i1})$ estimated using inverse probability weights, $w_i = 1/\widehat{\pi}_{i2}$, with the same terms that were used for modeling $\widehat{\pi}_{i2}$ (although this is not essential).

3. Fitting a linear regression model for completers (subjects with non-missing outcomes at Time 3) $m_3\left(x_i, y_{i1}, y_{i2}^*\right)$ with weights, $w_i = 1/\hat{\pi}_{i3}$, and with the same terms that were used for modeling $\hat{\pi}_{i3}$. Missing outcomes for changes from baseline to Time 2 were replaced with predicted values from the previous step (which is emphasized in the notation with an asterisk, y_{i2}^*).

4. Finally, we use predicted values $\hat{m}_{i3}, i = 1, ..., N$ obtained in the previous step as "new data" to fit an ANCOVA model with baseline score and treatment indicator as covariates.

5. The point estimate for the treatment contrast from the model in Step 4 is the final estimate of treatment effect at Time 3 (shown in Table 21.1), $\hat{\theta}_{aipw}$. The standard errors and p-values reported by ANCOVA (on the basis of full sample of N subjects) are severely biased downward because they do not account for the estimation done in previous steps, in particular the uncertainty due to imputation of missing values.

We obtain valid standard errors and confidence intervals by bootstrap (shown in Table 21.1).

Each bootstrap sample of size N is formed by sampling with replacement subject records (rows) of the observed $N \times p$ data matrix. Any subject irrespective of his or her missingness pattern has the same probability $1/N$ to be selected in each draw and may appear multiple times in each bootstrap data set. We form 2000 bootstrap data sets, and each gives rise to the same analysis Steps 1–5 as were applied to the observed data (including re-fitting models for computing weights) resulting in point estimates $\hat{\theta}_{aipw}^{(b)}, b = 1, ..., 2000$, which is the bootstrap distribution of our AIPW estimate.

A valid estimate of standard error is obtained as the standard deviation of the bootstrap distribution. We construct a 95% confidence interval using the

TABLE 21.1

Estimating Treatment Contrast and Least-Squares Means Using a Doubly Robust AIPW Method for Completers

		Complete Data			Incomplete Data (AIPWCC)		
Treatment	Time	LSMEANS	SE	95% CI	LSMEANS	SE	95% CI
1	3	−9.86	1.05	(−11.98, −7.75)	−9.70	1.11	(−11.62, −7.17)
2	3	−13.26	1.05	(−15.37, −11.14)	−12.58	1.26	(−15.19, −10.23)
Endpoint Treatment Difference		3.39	1.49	(0.39, 6.39)	2.88	1.71	(0.18, 6.82)

Note: Bootstrap-based confidence intervals and standard errors.

percentile method; that is, by taking as confidence limits the 2.5% and 97.5% percentile points of the bootstrap distribution. Other more sophisticated methods for bootstrap intervals can be used (e.g., the BCa, bias-corrected and accelerated bootstrap intervals, available in SAS %BOOTCI macro and R package *boot*).

Because the example data set is rather small, in some bootstrap samples, modeling probability of discontinuation via logistic regression may be challenging because of quasi-complete or even complete separation making the maximum likelihood estimation of parameters not possible. In many situations, this can be fixed using a variant of penalized regression known as Firth's penalized likelihood (Firth, 1993) (available in SAS PROC LOGISTIC). A small number of particularly unfortunate bootstrap samples (e.g., with no missing values) can be skipped.

21.5 Code Fragments

Here we implemented a simple version of AIPW estimator similar in spirit to Vansteelandt et al. (2010). To obtain bootstrap-based standard errors, we used %BOOT and %BOOTCI macros available at http://support.sas.com/kb/24/982.html. These macros require that the computation of estimates subjected to bootstrapping was implemented by the user in a separate macro called %ANALYZE. To facilitate "by-processing" when computing bootstrap distribution of estimates, %BYSTMT macro is called within each procedure, and data step inside %ANALYZE macro. The code may not be most efficient and specific to the data set at hand but easy to follow and (as we hope) generalize to other examples the reader may encounter.

Code Fragment 21.1   SAS code for augmenting IPW

```
%macro analyze(data= , out= );
/* predicting prob of dropping at vis 2 */
   ODS LISTING CLOSE;
   PROC LOGISTIC DATA=&DATA DESC;
    %BYSTMT;
   MODEL R2 =TREAT BASVAL YOBS1 TREAT*YOBS1 TREAT*BASVAL/FIRTH
MAXITER=100 ;
   OUTPUT OUT =PREDR2 P=PROB_V2;
   RUN;


/* predicting prob of dropping at vis 3 */
   PROC LOGISTIC DATA=PREDR2 DESC;
    %BYSTMT;
   MODEL R3 =TREAT BASVAL YOBS1 YOBS2 TREAT*YOBS1 TREAT*YOBS2
     TREAT*BASVAL/FIRTH MAXITER=100;
   OUTPUT OUT =PREDR3 P=PROB_V3;
   RUN;
```

```
/* computing weights at vis 2 and 3 */
DATA WEIGHTS_OBS ;
 SET  PREDR3;
 %BYSTMT;
 IF R2 =0 THEN DO;
   WEIGHT2=1/(1-PROB_V2);
 END;
 IF R3 =0 THEN DO;
   WEIGHT3=1/(1-PROB_V2)*1/(1-PROB_V3);
 END;
RUN;

/* predict missing outcomes for vis 2 using IPW model*/
    PROC GENMOD DATA=WEIGHTS_OBS ;
     %BYSTMT;
     MODEL YOBS2= TREAT BASVAL YOBS1 TREAT*BASVAL
     TREAT*YOBS1;
     WEIGHT WEIGHT2;
     OUTPUT OUT=PREDY2 PRED=PRED_Y2;
    RUN;

    /* replace missing values for vis 2 with predicted */
    DATA PREDY2;
     SET PREDY2 ;
   %BYSTMT;
     IF YOBS2= . THEN YOBS2=PRED_Y2;
    RUN;

    /* predict outcomes for vis 3 using IPW model */
    PROC GENMOD DATA=PREDY2;
     %BYSTMT;
    MODEL YOBS3= TREAT BASVAL YOBS1 YOBS2 TREAT*BASVAL
TREAT*YOBS1 TREAT*YOBS2;
     WEIGHT WEIGHT3;
     OUTPUT OUT=PREDY3 PRED=PRED_Y3;
    RUN;

    ODS OUTPUT LSMEANS=LSDR1
              ESTIMATES=ESDR1;
    PROC GENMOD DATA=PREDY3;
     %BYSTMT;
     CLASS TREAT;
     MODEL PRED_Y3= BASVAL TREAT;
     LSMEANS TREAT;
     ESTIMATE "DR ESTIMATE OF TRT" TREAT -1 1;
    RUN;

  PROC TRANSPOSE DATA=LSDR1  OUT= LSMEANS_TRANSP
(DROP=_NAME_)  PREFIX=LSMEANS;
```

```
       %BYSTMT;
     VAR ESTIMATE;
     ID   TREAT;
     RUN;

       DATA &OUT;
       MERGE ESDR1 (KEEP=MEANESTIMATE &BY RENAME=
          (MEANESTIMATE =TRTDIFF3))
          LSMEANS_TRANSP;
     %BYSTMT;
       RUN;
       ODS LISTING;
  %MEND;

/* construct analysis data with single record per subject */
 PROC TRANSPOSE DATA=DATALONG OUT= DATAWIDE (DROP=_NAME_)
PREFIX=YOBS;
       BY TRT SUBJECT BASVAL;
       VAR CHGDROP;
       ID   TIME;
 RUN;


DATA FOR_ANALYSIS;
 SET DATAWIDE;
 R2=(YOBS2=.);
 R3=(YOBS3=.);
 TREAT=(TRT="2");
RUN;

/* applying analyze macro to observed data */
%LET BY=;
 %ANALYZE(DATA=FOR_ANAL, OUT=OUT_EST);

 %BOOT(DATA=FOR_ANAL, ALPHA=.05, SAMPLES=2000, RANDOM=123,
STAT = TRTDIFF3 LSMEANS0 LSMEANS1, CHART=0, BIASCORR=1);
 %BOOTCI(METHOD=PERCENTILE);
```

21.6 Summary

One motivation for doubly robust methods is to improve the efficiency of wGEE by augmenting the estimating equations with the predicted distribution of the unobserved data given the observed data. This augmentation also introduces the property of double robustness.

Efficient IPW estimators require three models: (1) the substantive (analysis) model that relates the outcome to explanatory variables and/or covariates of interest, (2) a model for the probability of observing the data (usually a logistic model of some form), and (3) a model for the joint distribution of the partially and fully observed data, which is compatible with the substantive model in (1).

With augmented wGEE if either model (2) or model (3) is wrong, but not both, the estimators in model (1) are still consistent. If model (1) is wrong, then estimates of all parameters will typically be inconsistent.

Doubly robust methods are more recent, with few rigorous evaluations in practical settings. Therefore, our understanding of how to best apply these methods in clinical trial scenarios is emerging.

22

Reference-Based Imputation

22.1 Introduction

As discussed in Chapter 11, reference-based imputation (RBI) is useful for addressing estimands that involve treatment switching. Examples of such scenarios include what would have happened if patients took placebo or some reference (rescue) treatment after discontinuing the initially randomized study drug. In reference-based methods, the common idea is to construct a principled set of imputations that model specific changes in treatment. After relevant intercurrent events (ICEs), values are imputed by making qualitative reference to another arm in the trial. In other words, imputed post-ICE values have the statistical behavior of otherwise similar patients in a reference group. Initial implementations of RBI used multiple imputation. However, likelihood-based and Bayesian analogs have been proposed. This chapter provides technical details on the various approaches.

The authors credit Frank Liu and James Roger as co-author through their contribution to the paper upon which this Chapter is based.

22.2 Multiple Imputation-Based Approach

To explain the various multiple imputation-based implementations of RBI, assume the following notation. Let

n = the number of subjects.

q = the number of treatments.

t = the number of visits (assessment times).

$i = 1,...,n$ the individual subject indicator.

$j = 1,...,q$ the treatment randomly assigned to the ith subject.

$k = 1,...,t$ the individual visit indicator.

$r = 1,...,q$ the reference treatment arm.

p = the visit at which subject i withdraws.

s = an index for the covariates.

The underlying model is the standard Gaussian repeated-measures model for quantitative data with parameters A_{jk} for treatment-by-time interaction and β for all the baseline covariates.

$$E[Y_{ik}] = A_{jk} + \sum_s X_{iks}\beta_s \qquad (22.1)$$

The subjects are independent and the vector Y_i for the ith subject has a multivariate normal distribution with covariance matrix R_j. If data were complete, this repeated-measures model could be fit with the following features.

- A treatment-by-time interaction to indicate the profile across time within each treatment arm, A_{jk}.
- Other covariates, which may or may not be crossed with time, as defined in the matrix X_i.
- An unstructured covariance matrix that may be shared across arms or may be separate for each arm.

The same model can be fit when some Y are missing by assuming missing at random (MAR) (see Chapter 15). Estimated parameters from this model are used to build predicted profiles for the unobserved values in a separate imputation model, one for each pattern of withdrawal. These profiles represent outcomes to expect in subjects after withdrawal. Missing values for an individual are imputed based on the profile of predicted means that matches their treatment and their time of withdrawal.

Three models are required in this process.

1. A parameter-estimation model, which can be a likelihood-based repeated measures model that assumes MAR. This is a marginal model, not a pattern-specific model.
2. The imputation model, in which a model for predicted values is built for each pattern of withdrawal using the parameters obtained from the parameter estimation model (1). How the parameters in these models link to the parameters estimated in (1) define the different reference-based methods.
3. An analysis model that analyses the complete data once for each draw (imputation). This is often simply a univariate analysis of variance (ANOVA) model.

The crucial missing not at random (MNAR) part of the process that addresses treatment switching is how the profile across time is defined for each pattern and treatment in the imputation model, based on the values of the parameters in the estimation model. The only difference between patterns is in the parameters of the treatment-by-time interaction, which are the profiles across time for each treatment arm. In the estimation model, there is a single set of $q \times t$ parameters A_{jk}, while in the imputation model, there is a set B_p of $q \times t$ parameters, one for each possible withdrawal visit ($p = 0,...,t$) where apart from the last visit $p = t$, the data are missing **after** the pth visit. When $p = 0$, all data for the subject are missing:

$$E_p\left[Y_{ik}\right] = B_{pjk} + \sum_s X_{iks}\beta_s \tag{22.2}$$

Different MNAR models are defined by specifying the relationship between B_p and A. For any withdrawal pattern p, the response vector Y splits into two parts: the observed values Y_1 prior to withdrawal and the unobserved values Y_2 that are imputed. For a single imputation (drawn from the posterior distribution of the parameters), the imputation model has a mean with two parts, μ_1 and μ_2, and a variance-covariance matrix with four components $\Sigma_{11}, \Sigma_{12}, \Sigma_{21}$ (which is the transpose of Σ_{12}) and Σ_{22}.

Imputed values are sampled by drawing from the distribution of Y_2 conditional upon the observed value of Y_1. This distribution is also multivariate normal with mean $\mu_2 + \Sigma_{21}\Sigma_{11}^{-1}(Y_1 - \mu_1)$ and variance-covariance matrix $\Sigma_{22} - \Sigma_{21}\Sigma_{11}^{-1}\Sigma_{12}$. In the model definition, A_{jp} acts as an intercept for each combination of treatment and time. Its value is thus dependent on the constraints applied in the definition of the covariate design matrix X_i. The subject's covariate offset $\Sigma_s X_{iks}\beta_s$ is the same whatever treatment j they receive but varies across times k. Calculate the average of these profiles across all the subjects $C_k = \frac{1}{n}\Sigma_{is} X_{iks}\beta_s$ and re-express the estimation model as

$$E\left[Y_{ik}\right] = A_{jk} + \sum_s X_{iks}\beta_s - \frac{1}{n}\sum_{is} X_{iks}\beta_s \tag{22.3}$$

and the imputation model as

$$E_p\left[Y_{ik}\right] = B_{pjk} + \sum_s X_{iks}\beta_s - \frac{1}{n}\sum_{is} X_{iks}\beta_s \tag{22.4}$$

In this way, A_{jk} is invariant under change to the constraints. In other words, A_{jk} is defined as the treatment-by-time intercept for a subject with a profile based on average values for covariate offset. It is the offset for an "average" subject. The different methods are defined in terms of these average sets of intercepts.

22.2.1 Missing at Random

Standard multiple imputation assuming MAR (no change in treatment) is implemented by:

$$B_{pjk} = A_{jk} \text{ for all patterns } p, j, \text{and } k. \tag{22.5}$$

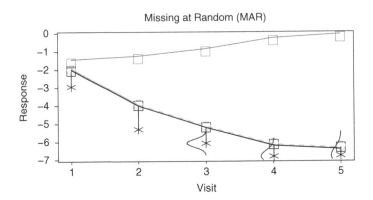

FIGURE 22.1
Illustration of multiple imputation based on MAR.

All the patterns have the same profile, so that, conditional on covariates, missingness is independent of the profile. This is equivalent to a likelihood-based repeated measures model and estimates a *de jure* estimand. The MAR approach is depicted graphically in Figure 22.1 (taken from Mallinckrodt and Lipkovich, 2017). Here the actual values for a subject discontinued from active treatment arm after visit 2 are depicted with "x"s placed below the line (representing the group mean trajectory) at visits 1 and 2. Imputed values are sampled from distributions depicted with normal curves (at visits 3,4,5) centered around conditional means that reflect the effect of "regression to the mean" and the fact the subject's observed values prior to withdrawal were better than the group's mean.

22.2.2 Copy Increment from Reference

The copy increment from reference method of RBI is implemented by

$$B_{pjk} = A_{jk} \text{ for } k \leq p$$

$$= A_{jp} + A_{rk} - A_{rp} \text{ for } k > p \tag{22.6}$$

All patterns have the same profile up to withdrawal (as estimated for this arm in the parameter estimation model), but after withdrawal, the pattern tracks parallel to the pattern for the reference arm. For the reference arm, this is equivalent to MAR. The assumption is that when a subject stops treatment, they continue to take advantage of their previous therapy, but from withdrawal onwards, they progress in the same way as the subjects in the reference arm (parallel to the profile for the reference arm). As such, it will usually answer a *de facto* question and be especially applicable for drugs that modify the progression of disease. The copy increment from reference (CIR) approach is depicted graphically in Figure 22.2 (taken from Mallinckrodt and Lipkovich, 2017).

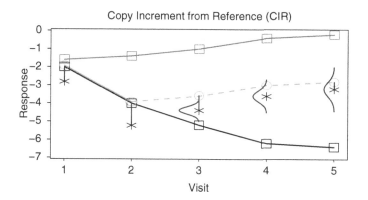

FIGURE 22.2
Illustration of copy increment from reference-based imputation.

22.2.3 Jump to Reference

The jump-to-reference (J2R) method of RBI is implemented by

$$
\begin{aligned}
B_{pjk} &= A_{jk} \text{ for } k \le p \\
&= A_{rk} \text{ for } k > p
\end{aligned}
\tag{22.7}
$$

All the patterns have the same profile up to withdrawal (which estimated in the parameter estimation model), but after withdrawal the profile jumps to the estimated profile for the reference arm. As for CIR, the reference arm profile is equivalent to MAR. The J2R approach is depicted graphically in Figure 22.3 (taken from Mallinckrodt and Lipkovich, 2017).

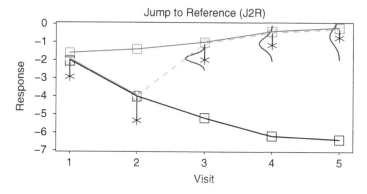

FIGURE 22.3
Illustration of jump to reference-based imputation.

22.2.4 Copy Reference

The copy increment from reference method of RBI is implemented by

$$B_{pjk} = A_{rk} \text{ for all } p, j, \text{ and } k. \tag{22.8}$$

All patterns, except patterns for those who do not withdraw, follow the whole profile estimated for the reference arm in the parameter estimation model, both before and after withdrawal. Consequently, this method is distinctly different from the rest of the methods in that the profile prior to withdrawal is that for the reference arm rather than that for the subjects' own arm. Hence the residuals $\left(Y_1 - \mu_1\right)$ are measured from mean μ_1 for the reference arm, rather than that for the subject's own arm in the formula $\mu_2 + \Sigma_{21}\Sigma_{11}^{-1}\left(Y_1 - \mu_1\right)$ for the conditional mean used in imputation. As for CIR and J2R, the reference arm profile for CR is equivalent to MAR. The CR approach is depicted graphically in Figure 22.4 (taken from Mallinckrodt and Lipkovich, 2017).

Commercially available software to implement RBI is limited. It is possible to implement the CR approach in SAS. However, J2R and CIR must be implemented using specialty coding. The Drug Information Association's (DIA) Scientific Working Group on missing data has SAS macros available at missingdata.org.uk, under the DIA working group tab, to implement a variety of RBI approaches, along with other specialty software tools.

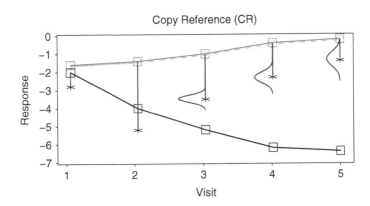

FIGURE 22.4
Illustration of copy reference-based imputation.

TABLE 22.1

Results from Copy Reference Analyses of the Small Example Data Set with Dropout

Method	Endpoint Contrast	Standard Error	*p*-Value	Deviation from MAR
MAR-Based MI	2.98	1.71	0.082	0.00
Copy reference	2.69	1.64	0.103	0.31

22.2.5 Example

The code fragment to implement CR in SAS for the small example data set with dropout is listed in Section 22.6 (Code Fragment 22.1). Results from that analysis are summarized in Table 22.1. The deviation from MAR, based on difference in treatment contrast from the MAR analysis, when applying CR was 0.31 (2.98 versus 2.69). This deviation is smaller than would usually be the case in actual scenarios because only two post-baseline visits had missing values, and these outcomes were strongly correlated.

22.3 Group Mean Imputation Using a Likelihood-Based Approach

Using the same notation as in the previous section, define a likelihood-based repeated measures model for the mean responses for patient $i(=1,...,n)$ at time $k(=1,...,t)$

$$E\left[Y_{ijk}\right] = A_{jk} + X_{ik}\beta_k \equiv \mu_{jk}, \tag{22.9}$$

where $j(=1,...,q)$ is the indicator for treatment group, X_{ik} is a collection of covariates to adjust the analysis model, which are centralized as in (22.3), β_k is the vector of corresponding coefficients, and A_{jk} is the treatment effect at time k. Also assume that the response vector $Y_{ij} = \left(Y_{ij1},...,Y_{ijt}\right)^T$ has mean $\mu_j = \left(\mu_{j1},...,\mu_{jt}\right)^T$ as defined in (22.9), an unstructured covariance matrix Σ, with the coefficients β_k and covariance matrix Σ the same across treatment groups. With centralized X_{ik}, the A_{jk} can be interpreted as the treatment effect at the mean $\bar{X}_k (=0)$ for group j at time k. Hence the treatment difference at time k can be constructed from the parameters of $\{A_{jk}, j = 1,..., q\}$. For example, in a trial with two treatment groups, say test drug ($k = 1$) and placebo ($k = 2$), then the treatment difference at the last time point is defined as $\theta_{12}^{MAR} = A_{1t} - A_{2t} = \mu_{1t} - \mu_{2t}$. The superscript MAR is used to indicate that this treatment difference is for a *de jure* estimand under MAR, which assesses the mean difference between group 1 and group 2 assuming all subjects adhered to study medication.

With no missing data, this treatment difference can be estimated from the repeated measures model or even a univariate analysis of covariance (ANCOVA) model for the observations $\{y_{ijt}, i = 1, \ldots, n\}$ with factors of treatment and baseline covariates. The results from this univariate ANCOVA model will be the same as that from the repeated measures model with unstructured modeling of covariance and time (time is class effect) in the repeated measures analysis that includes the same effects as the ANCOVA model and their interactions with time.

With missing data, the parameters can be estimated from the repeated measures model that implicitly assumes Y_{ij} follow a multivariate normal distribution. Under MAR, the parameters in (22.9) and the treatment differences can be estimated using likelihoods constructed from observed data (the missing data can be ignored). This analysis is estimating what would happen when all subjects adhere.

Reference-based controlled mean imputation addresses a different estimand – the difference between treatments regardless of adherence and without initiation of rescue medication. As described in the previous section, the mean profile for response Y_{ik} is defined as

$$E\left[Y_{ik}\right] = B_{pjk} + X_{ik}\beta_k \tag{22.10}$$

where $p = 1, \ldots, t$ is the last time point with an observation for subject i. So $\{Y_{i(p+1)}, \ldots, Y_{it}\}$ are missing. The values B_{pjk} are specified based on the group mean reference-based controlled imputation strategy.

Assumptions are made for the mean parameters B_{pjk} under the group mean reference-based controlled imputation strategy. For the reference group (e.g., placebo), it is assumed that $B_{prk} = A_{rk}$ across time points, that is, the means for subjects who do not adhere in the test group are the same as those who stay in the study and take placebo. This is a reasonable assumption if subjects have no lasting benefit from a test drug if they stop taking it – that it is a symptomatic treatment rather than a disease-modifying treatment. Using notation, the mean of subjects who were not in the reference group and who dropped out at time p is based on the reference group. Specifically,

$$\text{when } k \leq p, \quad \text{when} \quad k > p$$

$$B_{pjk} = \begin{cases} A_{jk}, & A_{jp} + A_{rk} - A_{rp}, & CIR \\ A_{jk}, & A_{rk} & , & J2R \\ A_{rk}, & A_{rk} & , & CR \end{cases} \tag{22.11}$$

To simplify notation, ignore the subject indicator i, and let $y_{jo} = (y_{j1}, \ldots, y_{jp})^T$ be the observed subvector for this subject and $Y_{jm} = (Y_{j(p+1)}, \ldots, Y_{jt})^T$ be the subvector for missing data, and

$$\mu_j = \begin{pmatrix} \mu_{jo} \\ \mu_{jm} \end{pmatrix}, \quad \Sigma = \begin{pmatrix} \Sigma_{oo} & \Sigma_{om} \\ \Sigma_{mo} & \Sigma_{mm} \end{pmatrix}$$

be split subvectors and block matrices with dimensions corresponding to the y_{jo} and Y_{jm}. Then the missing data vector will be imputed from a conditional distribution

$$Y_{jm} \mid y_{jo}, X, \mu_j, \Sigma \sim N\left(\mu_{jm} + \Sigma_{mo}\Sigma_{oo}^{-1}\left(y_{jo} - \mu_{jo}\right), \Sigma_{mm} - \Sigma_{mo}\Sigma_{oo}^{-1}\Sigma_{om}\right).$$

Hence the unconditional mean subvector for the missing data (evaluated at $X = 0$) after applying the reference-based controlled imputation is

$$E\left[E\left(Y_{jm} \mid y_{jo}, \overline{X} = 0\right)\right] = \begin{cases} A_{jp} + A_{rp} - A_{rp}, & \text{CIR} \\ A_{rp}, & \text{J2R} \\ A_{rp} + \Sigma_{mo}\Sigma_{oo}^{-1}\left(A_{jo} - A_{ro}\right), & \text{CR} \end{cases} \quad (22.12)$$

where $A_{rp} = \left(A_{r(p+1)},\ldots,A_{rt}\right)^T$, $A_{jo} = \left(A_{j1},\ldots,A_{jp}\right)^T$ and $A_{ro} = \left(A_{r1},\ldots,A_{rp}\right)^T$, $p = 1,\ldots,t-1$. For completers, $E\left(Y_j \mid X = 0\right) = A_j = \left(A_{j1},\ldots,A_{jt}\right)^T$

Suppose f_{jp} is the proportion of subjects who have data in group j at time p, and let $f_{jt} = 1 - \sum_{p=1}^{t-1} f_{jp}$ be the proportion of completers for group j. Then the overall mean at the last time point t for treatment j can be expressed as

$$\theta_{jt} = \begin{cases} \sum_{p=1}^{t} f_{jp}\left(A_{jp} + A_{rt} - A_{rp}\right), & \text{CIR} \\ f_{jt}A_{jt} + \left(1 - f_{jt}\right)A_{rt}, & \text{J2R} \\ f_{jt}A_{jt} + \sum_{p=1}^{t-1} f_{jp}\left(A_{rt} + \left[\Sigma_{mo}\Sigma_{oo}^{-1}\left(A_{jo} - A_{ro}\right)\right]_t\right), & \text{CR} \end{cases} \quad (22.13)$$

where $\left[\Sigma_{mo}\Sigma_{oo}^{-1}\left(A_{jp} - A_{rp}\right)\right]_{jt}$ is the last element (at time t) of the subvector. Therefore, the treatment effect under reference-based group mean imputation can be expressed as a linear combination of the parameters of $\{A_{jk}\}$ in (22.9) and the proportions of subjects in each missing data pattern (i.e., time of dropout).

The treatment difference between groups can be constructed from the overall means as given in (22.13). In general, it will be a function of Σ, $f_j = \left(f_{j1},\ldots,f_{jt}\right)^T$ and $A_j = \left(A_{j1},\ldots,A_{jt}\right)^T$, $j = 1,\ldots,q$. One of the treatment groups, commonly the placebo group in placebo-controlled trials, will be used as the reference group. The treatment difference between group j and reference group r is $\theta_{jr}^{RBI} = \theta_{jt} - \theta_{rt}$. The superscript RBI indicates use of RBI.

With (22.13), θ_{jr}^{RBI} can be estimated using a likelihood approach. Specifically, the parameters of Σ and $\{A_j, j = 1,\ldots,q\}$ can be obtained from the repeated measures model using likelihood estimation. Suppose $\hat{f}_j = (\hat{f}_{j1}, \ldots, \hat{f}_{jt})^T$ is the observed proportions of dropouts or completers, $\hat{\Sigma}$ and $\{\hat{A}_j, j = 1, \ldots, q\}$ are estimated from the repeated measures analysis. The point estimate and its standard error for a treatment difference are obtained treating $\hat{\Sigma}$ as fixed, with the treatment difference θ_{jr}^{RBI} estimated as a linear combination of $\hat{A}_j - \hat{A}_r$.

$$
\hat{\theta}_{jr}^{RBI} = \begin{cases} \displaystyle\sum_{p=1}^{t} \hat{f}_{jp} \left(\hat{A}_{jp} - \hat{A}_{rp} \right), & \text{CIR} \\[2ex] \hat{f}_{jt} \left(\hat{A}_{jt} - \hat{A}_{rt} \right), & \text{J2R} \\[2ex] \hat{f}_{jt} \left(\hat{A}_{jt} - \hat{A}_{rt} \right) + \displaystyle\sum_{p=1}^{t-1} \hat{f}_{jp} \left[\hat{\Sigma}_{mo} \hat{\Sigma}_{oo}^{-1} \left(\hat{A}_{jo} - \hat{A}_{ro} \right) \right]_t, & \text{CR} \end{cases}
$$
(22.14)

To obtain its variance, use the formula

$$
var\left(\hat{\theta}_{jr}^{RBI} \right) = E\left[var\left(\hat{\theta}_{jr}^{RBI} \mid \hat{f}_j \right) \right] + var\left[E\left(\hat{\theta}_{jr}^{RBI} \mid \hat{f}_j \right) \right].
$$
(22.15)

The conditional variance of the first term given $\hat{f}_j$ can be derived from the variance estimates from the repeated measures model (e.g., using the covariance estimates of the LSMEAN differences from the SAS PROC MIXED analysis). The second term can be calculated using the point estimates of $\hat{\theta}^{RBI}$ and $var\left(\hat{f}_j \right) = (v_{ij})$, with $v_{ik} = \frac{f_{ji}(1 - f_{jk})}{n_j}$ for $i = k$, $-\frac{f_{ji}f_{ji}}{n_j}$ for $i \neq k$, where n_j is the sample size for group j.

As an example, derive the variance for J2R in a trial with two treatment groups, say test drug ($k = 1$) and placebo ($k = 2$). The treatment difference at the last time point with J2R is defined as $\theta_{12}^{J2R} = f_{1t}(A_{1t} - A_{2t})$, where f_{1t} is the proportion of completers in the test drug group. Suppose $\hat{\theta}_{12}^{MAR} = \hat{A}_{1t} - \hat{A}_{2t}$ is the likelihood-based repeated measures estimate for the treatment difference with variance estimate $\hat{v}^{MAR}$. Then the treatment difference under J2R is estimated as $\hat{\theta}_{12}^{J2R} = \hat{f}_{1t} \hat{\theta}_{12}^{MAR}$, and its variance estimate is

$$
var\left(\hat{\theta}_{12}^{J2R} \right) = E\left[var\left(\hat{\theta}_{12}^{J2R} \mid \hat{f}_{1t} \right) \right] + var\left[E\left(\hat{\theta}_{12}^{J2R} \mid \hat{f}_{1t} \right) \right]
$$

$$
= \hat{v}^{MAR} E\left[\hat{f}_{1t}^2 \right] + \left(\hat{\theta}_{12}^{MAR} \right)^2 var\left[\hat{f}_{1t} \right]
$$
(22.16)

$$
= \hat{v}^{MAR} \hat{f}_{1t}^2 + \left[\hat{v}^{MAR} + \left(\hat{\theta}_{12}^{MAR} \right)^2 \right] \frac{\hat{f}_{1t}(1 - \hat{f}_{1t})}{n_1}.
$$

The estimates and inference for the RBI parameters are based on the parameters in (22.9), which can be estimated using likelihood-based repeated measures under an MAR assumption. The MAR assumption here is used to define the latent mean profiles of (22.9). The treatment difference as constructed from contrasting $\{A_{jt}, j = 1,\ldots, q\}$ is the estimand of interest, which is based on a special pattern-mixture model as specified by the RBI strategy. It is assumed that the missing data for the reference group are MAR. However, the distribution of values after withdrawal is different from those who stay in the study for the nonreference groups. Therefore, the missing data mechanism for RBI is MNAR.

Importantly, the MNAR approach for RBI is different from assuming MNAR for equation (22.9) to allow missing data to depend on the unobserved data. The latter causes the parameters of $\{A_{jk}\}$ to be undefined based on the observed data without knowing the missing data mechanism. If the true parameters (always unknown) in (22.9) are from MNAR, the RBI analysis will still use MAR to define the latent parameters for (22.9) based on all observed data. These latent parameters will be different from the "true" parameters under MNAR. Therefore, under this MNAR scenario, the RBI analysis will be biased and may inflate type-I error as shown in simulations by Liu and Pang (2016). Thus, the RBI methods are specific pattern-mixture models under the principle of reference-based group mean controlled imputation that are defined using the latent parameters obtained under MAR.

22.4 Bayesian-Based Approach

Bayesian methods are a natural approach to modeling the uncertainty from missing data in longitudinal trials (Daniels and Hogan, 2008). Bayesian methods have also been applied to RBI (Liu and Peng, 2016). With advancements in computation and statistical software, simulation-based methods such as Markov Chain Monte Carlo (MCMC) sampling become feasible for analysis of longitudinal clinical trials. Computationally, MCMC generates a series of samples for model parameters to approximate their equilibrium posterior distributions, and then inference is drawn based on these distributions.

Bayesian methods start with specification of prior distributions and initial values for model parameters. In applications with no specific prior information, noninformative or flat priors are used, and statistical inference is similar to likelihood-based methods. The samples are updated iteratively using direct sampling when explicit posterior distributions are available, such as under conjugacy of priors or using data augmentation methods such as Metropolis or Metropolis-Hastings algorithms for complicated distributions (Tanner and Wong, 1987).

First, consider a Bayesian model for repeated measures in a longitudinal study with no missing data. Assume $Y_{ij} = \left(Y_{ij1}, \ldots, Y_{ijt}\right)^T$ ~ MultiNormal$\left(\mu_j, \Sigma\right)$, where mean vector μ_j is defined in (22.15) with parameters of $\{A_{jk}, j = 1, \ldots, q, k = 1, \ldots, t\}$ and $\{\beta_k, k = 1, \ldots, t\}$. The likelihood function can be written as

$$L(A_{jk}, \beta_k, \Sigma \mid Y_{ijk}) = \prod_{i=1}^{n} f(Y_{ij} \mid \mu_j, \Sigma),$$

where $f(Y_{ij} \mid \mu_j, \Sigma)$ is the density function of a multivariate normal with mean μ_j and covariance matrix Σ. Suppose $g_1\left(A_{jk}, \beta_k, j = 1, \ldots, q; k = 1, \ldots, t\right)$ is the prior for parameters in the mean μ_j, and $g_2(\Sigma)$ is a prior for covariance. Then the posterior distribution for $\{A_{jk}, \beta_k, \Sigma\}$ will be

$$f(A_{jk}, \beta_k, \Sigma \mid Y_{ijk}) = g_1\left(A_{jk}, \beta_k\right) g_2\left(\Sigma\right) \prod_{i=1}^{n} f(Y_{ij} \mid \mu_j, \Sigma) \qquad (22.17)$$

Without specific prior information for the parameters, noninformative priors may be used. For example, $g_1\left(A_{jk}, \beta_k\right) \propto 1$, and Σ ~ invWishart(I, t), an inverse Wishart distribution with dimension t, and I as an identity matrix of the same dimension. Then the posterior distribution samples can be obtained through MCMC for (22.17) using software packages such as SAS PROC MCMC and STAN. The posterior samples for any function of these parameters, for example, the treatment difference between groups at the last time point t, $\theta_{jl}^{MAR} = A_{jt} - A_{lt}$, can be obtained. Statistical estimates such as mean and standard deviation and credible intervals can be obtained from these posterior samples.

In longitudinal studies with missing data, the missing values are treated as parameters in Bayesian analysis and directly sampled from the conditional distribution. For a repeated measures model that assumes MAR, simplify the notations by ignoring the subject indicator i, and let $y_{jo} = (y_{j1}, \ldots, y_{jp})^T$ be the observed subvector for a subject, and $Y_{jm} = (Y_{j(p+1)}, \ldots, Y_{jt})^T$ be the subvector for missing data, and

$$\mu_j = \begin{pmatrix} \mu_{jo} \\ \mu_{jm} \end{pmatrix}, \quad \Sigma = \begin{pmatrix} \Sigma_{oo} & \Sigma_{om} \\ \Sigma_{mo} & \Sigma_{mm} \end{pmatrix}$$

are subvectors and block matrices with dimensions corresponding to the y_{jo} and Y_{jm}. Using the Bayesian approach, the missing data is directly sampled from

$$Y_{jm} \mid (y_{jo}, X_j, A_{jk}, \beta_k, \Sigma) \sim N\left(\mu_{jm} + \Sigma_{mo}\Sigma_{oo}^{-1}\left(y_{jo} - \mu_{jo}\right), \Sigma_{mm} - \Sigma_{mo}\Sigma_{oo}^{-1}\Sigma_{om}\right).$$

$$(22.18)$$

Then the samples of model parameters can be drawn from the posterior distribution (22.17) based on combined observed data y_{jo} and sampled missing data Y_{jm}. Starting with the initial parameter values, the sampling process is carried out iteratively as follows:

1. Sampling $Y_{jm}|(y_{jo}, X_j, A_{jk}, \beta_k, \Sigma)$ as in (22.18) across patients in the study,

2. Sampling $A_{jk}, \beta_k, \Sigma | (y_{jo}, Y_{jm})$ as in (22.17),

3. Calculating parameter of interest, for example, $\theta_{12}^{MAR} = A_{1t} - A_{2t}$.

Statistical estimates such as mean and standard deviation and credible intervals for θ_{12}^{MAR} can be obtained from the posterior samples for statistical inference.

To reduce the impact of initial values on the posterior distribution samples, it is common practice to discard a portion of the MCMC samples (called burn-in) and use the samples after the burn-in for statistical inference. For complicated distributions, the mixing (or convergence) of the MCMC samples may be slow. It is therefore important to check the convergence of MCMC samples for all parameters through reviewing trace plots or checking test statistics as provided in Bayesian computation software, for example, SAS PROC MCMC (SAS/STAT 14.3).

The Bayesian approach for repeated measures requires MAR. Specifically, the direct sampling of the missing data (22.18) is based on the same conditional distribution as those who complete the study. With noninformative prior distributions, the inference from Bayesian analyses is similar to that from likelihood-based methods using mixed models (Liu and Pang, 2016).

For RBI approaches, the treatment difference can be expressed as a function of $\{A_{jk}, \beta_k, \Sigma\}$ and $f_j = (f_{j1}, \ldots, f_{jt})^T$ as given in (22.14). Bayesian analyses treat these unknown parameters as random variables. To reflect the randomness of the proportions of missing data patterns, sample the proportion of missing data pattern $f_j = (f_{j1}, \ldots, f_{jt})^T$ from a Dirichlet distribution with probability density

$$p(f_{j1}, \ldots, f_{jt} | n_j) \propto \prod_{k=1}^{t} f_{jk}^{n_{jk}-1}, \tag{22.19}$$

where $f_{jk} > 0$ is the proportion of dropout for treatment group j at time $k, k = 1, \ldots, t-1$ or completers for $k = t$; $\sum_{k=1}^{t} f_{jk} = 1$; $n_j = (n_{j1}, \ldots, n_{jt})^T$ and n_{jk} is the observed number of subjects in treatment group j who dropped out at time $k, k = 1, \ldots, t-1$ or completers for $k = t$. Specifically, starting with the initial values of the parameters, the sampling process is carried out iteratively as follows,

1. Sampling $Y_{jm}|(y_{jo}, X_j, A_{jk}, \beta_k, \Sigma)$ as in (22.18) across patients in the study

2. Sampling $A_{jk}, \beta_k, \Sigma | (y_{jo}, Y_{jm})$ as in (22.17)

3. Sampling $\{f_{j1},\ldots,f_{jt}\}\,|\,n$ as in (22.19)

4. Calculating parameter of interest, for example, $\theta_{12}^{J2R} = f_{1t}(A_{1t} - A_{2t})$

Statistical estimates such as mean and standard deviation and credible intervals for θ_{12}^{J2R} can be obtained from the posterior samples for statistical inference.

Simulations by Liu and Pang (2016) also showed that the point estimates, standard deviations, type-1 error rate, and power from the posterior samples for treatment contrasts under reference-based group mean imputation strategies were similar to those obtained from approximation using the likelihood approach as described in formulas (22.15) and (22.16). In this Bayesian approach, the variation of the dropout proportion f_j is incorporated into the posterior distribution through sampling the f_j from the Dirichlet distribution.

22.5 Considerations for the Variance of Reference-Based Estimators

Important considerations are embedded within the computations for the variance of the various reference-based group mean estimators. In reference-based methods using multiple imputation, the variance and standard error of treatment contrasts are determined using the same method as in MAR-based multiple imputation. Rubin's rules are used to combine the between- and within-imputation variances. This approach explicitly imputes missing values for each subject in the experimental group using a model developed from the reference group. It explicitly accounts for the uncertainty (variability) in the imputed values when determining the variance and standard error of treatment contrasts while maintaining the statistical independence of the reference and experimental arms. However, this MI-based variance may overestimate the true variability compared to the empirical variance from simulations.

The likelihood- and Bayesian-based approaches to group mean reference-based imputation presented above take a different approach. Rather than imputing individual values with uncertainty, these methods *define* the mean of subjects in the experimental group who discontinued to be equal to the mean of the control group. Because values are defined without uncertainty rather than imputed with uncertainty, the variance and standard errors of treatment contrasts reflect no uncertainty in the imputation. Therefore, group mean-based approaches generally yield smaller standard errors than the corresponding MI-based methods. Moreover, the standard errors will also be smaller than in the corresponding likelihood-based repeated measures analysis.

Intuitively, the standard error when imputing data should not be smaller than the standard error based on the observed data because imputation does not create information. In fact, equation (22.15) shows that in group mean imputation if every subject dropped out, the point estimate and the variance would both be zero. However, the approach used to obtain the variance of treatment contrasts from group mean-based methods does not require explicit data imputation and can be valid when the assumptions of the reference-based approach are justifiable.

Specifically, if subjects who discontinue an experimental drug are defined to have no benefit from it, the outcome is known – zero benefit. Group mean imputation uses the mean of the reference group to ascribe zero benefit to experimental group dropouts. This approach of defining the outcome is consistent with composite approaches to dealing with ICEs wherein dropout is considered an outcome – here that outcome is zero benefit. This variance shrinkage is also seen in other methods for obtaining the variance and standard error of treatment contrasts, such as bootstrapping. Simulation studies indicate that the average of these likelihood- and Bayesian-based variance estimates is close to the empirical variance from simulations (Liu, and Peng, 2016; Mehrotra, Liu, and Permutt, 2017).

Alternative approaches for estimating the variance in group mean reference-based imputation can incorporate uncertainty of imputation. To see how this can be done in principle, consider a simple example of a study with an outcome measured at a single time (rather than longitudinally) (adopted from Mehrotra, Liu, and Permutt, 2017). The goal is to estimate the effect of treatment in the presence of rescue that assumes a counterfactual outcome that patients who meet rescue criterion on either the active treatment or placebo arm would not (contrary to the fact) get access to any treatment (i.e., switch to placebo arm). To this end, a group-imputation strategy from the placebo arm was implemented and a variance estimator derived.

To fix ideas, let $T = 0,1$ denote placebo and active treatment arms, respectively, the ICE causing rescue be $S = 0,1$ (0 means a meeting rescue condition), $f_j = Pr(S = 1 | T = j), j = 0,1$. Further, let $\mu_{(1)|S=1}$ be the expected conditional outcome for patients initially randomized to active treatment who did not meet the rescue criterion and continued treatment, and $\mu_{(1,0)|S=0}$ be the expected outcome for those who met the rescue condition and subsequently stopped receiving treatment. Similarly $\mu_{(0)|S=1}$ and $\mu_{(0,0)|S=0}$ are defined for patients who were randomized to placebo. Note that our quantity of interest is

$$\delta = \mu_{(1)|S=1} f_1 + \mu_{(1,0)|S=1}(1 - f_1) - \left[\mu_{(0)|S=1} f_0 + \mu_{(0,0)|S=0}(1 - f_0) \right].$$

Note that interest may be in the estimate $\hat{\delta}$ either considering δ an estimand on its own or as a sensitivity analysis for a hypothetical strategy that assumes patients who meet discontinuation criteria would have remained on assigned treatment. For the rest of this discussion, we make a simplifying assumption that for placebo patients expected outcomes in absence of treatment are the same whether they met or did not meet rescue, that is, $\mu_{(0)|S=1} = \mu_{(0,0)|S=0}$. Therefore the target is,

$$\delta = \mu_{(1)|S=1}f_1 + \mu_{(1,0)|S=0}(1 - f_1) - \mu_{(0)},$$

where $\mu_{(0)}$ is the expected unconditional mean for placebo patients.

Now, it is tempting to also assume that $\mu_{(1,0)|S=0} = \mu_{(0)}$, that is patients in the treatment group who met rescue have the same expected outcome as those who were randomized to placebo such that $\hat{\mu}_{(0)}$ estimated from the placebo arm is the estimate for both groups. We can then proceed with the "plug-in" group imputation estimate, $\hat{\delta}_1 = \hat{f}_1(\hat{\mu}_{(1)|S=1} - \hat{\mu}_{(0)})$, where $\hat{f}_1$ is the proportion of patients in the treatment group who did not meet rescue, $\hat{\mu}_{(1)|S=1}$, estimated from corresponding observed outcomes, and $\hat{\mu}_{(0)}$ is the placebo mean. The variance of $\hat{\delta}_1$ can be estimated using simple formula for unconditional variance (Mehrotra, Liu, and Permutt, 2017) as

$$\widehat{var}\left(\hat{\delta}_1\right) = \left[\frac{\hat{f}_1\left(1-\hat{f}_1\right)}{n} + \hat{f}_1^{\,2}\right]\left(\widehat{var}\left(\hat{\mu}_{(1)|S=1}\right) + \widehat{var}\left(\hat{\mu}_{(0)}\right)\right)$$

$$+ \frac{\hat{f}_1\left(1-\hat{f}_1\right)}{n}\left(\hat{\mu}_{(1)|S=1} - \hat{\mu}_{(0)}\right)^2,$$

where n is the sample size in the treatment arm. As explained before, this estimator $\hat{\delta}_1$ shrinks the treatment effect toward zero and is zero with certainty if all patients meet rescue $\left(\hat{f}_1 = 0\right)$. Then both $\hat{\delta}_1$ and its variance are exactly 0.

However, if we do NOT assume $\mu_{(1,0)|S=0} = \mu_{(0)}$, but rather that our estimation strategy is mimicking a hypothetical trial where patients who met rescue are left untreated, thereby yielding estimates $\hat{\mu}_{(1,0)|S=0}$ and $\hat{\mu}_{(0)}$ from independent groups of patients. Consequently, we would obtain an estimate

$$\hat{\delta}_2 = \hat{\mu}_{(1)|S=1}\hat{f}_1 + \hat{\mu}_{(1,0)|S=0}(1 - \hat{f}_1) - \hat{\mu}_{(0)}.$$

The variance of $\hat{\delta}_2$ can be easily obtained. To keep things general, we include covariance terms for $\hat{\mu}_{(1,0)|S=0}$ and $\hat{\mu}_{(0)}$, assuming they are not necessarily independent. Then,

$$\widehat{var}\left(\hat{\delta}_2\right) = \frac{\hat{f}_1\left(1-\hat{f}_1\right)}{n}\left(\hat{\mu}_{(1)|S=1} - \hat{\mu}_{(1,0)|S=0}\right)^2 + \left[\frac{\hat{f}_1\left(1-\hat{f}_1\right)}{n} + \hat{f}_1^{\,2}\right]\widehat{var}\left(\hat{\mu}_{(1)|S=1}\right)$$

$$+ \left[\frac{\hat{f}_1\left(1-\hat{f}_1\right)}{n} + (1-\hat{f}_1^{\,2})\right]\widehat{var}\left(\hat{\mu}_{(1,0)|S=0}\right) - 2\left(1-\hat{f}_1\right)\widehat{cov}\left(\hat{\mu}_{(1,0)|S=0}, \hat{\mu}_{(0)}\right)$$

$$+ \widehat{var}\left(\hat{\mu}_{(0)}\right).$$

Since in practice we do NOT have independent data to estimate $\hat{\mu}_{(1,0)|S=0}$, we end up with a sensitivity analysis by assuming that $\hat{\mu}_{(1,0)|S=0} = \hat{\mu}_{(0)} + c$, where c is a constant that quantifies how much worse or better the outcomes for hypothetical data can be compared to outcomes from those available in the placebo arm. Here we will assume $c = 0$. However, the key element is to keep and manipulate the $\widehat{cov}(\hat{\mu}_{(1,0)|S=0}, \hat{\mu}_{(0)})$ term in the equation so as to enforce the variability that is expected if the hypothetical trial was actually conducted.

There are several alternatives. One is to assume that the hypothetical outcomes are the same as observed placebo outcomes, that is $\widehat{cov}(\hat{\mu}_{(1,0)|S=0}, \hat{\mu}_{(0)}) = \widehat{var}(\hat{\mu}_{(0)})$, which results in the variance estimate $\widehat{var}(\hat{\delta}_1)$ from Mehrotra, Liu, and Permutt (2017). Another approach is to assume independent data, with $\widehat{cov}(\hat{\mu}_{(1,0)|S=0}, \hat{\mu}_{(0)}) = 0$. This results in an alternative estimator,

$$\widehat{var}\left(\hat{\delta}_2\right) = \left[\frac{\hat{f}_1\left(1-\hat{f}_1\right)}{n} + \hat{f}_1^{\,2}\right]\left(\widehat{var}\left(\hat{\mu}_{(1)|S=1}\right) + \widehat{var}\left(\hat{\mu}_{(0)}\right)\right)$$

$$+ \frac{\hat{f}_1\left(1-\hat{f}_1\right)}{n}\left(\hat{\mu}_{(1)|S=1} - \hat{\mu}_{(0)}\right)^2 + 2\left(1-\hat{f}_1\right)\widehat{var}\left(\hat{\mu}_{(0)}\right).$$

Therefore, $\widehat{var}(\hat{\delta}_2) = \widehat{var}(\hat{\delta}_1) + 2(1-\hat{f}_1)\widehat{var}(\hat{\mu}_{(0)})$, where the additional term $2(1-\hat{f}_1)\widehat{var}(\hat{\mu}_{(0)})$ penalizes for not having observed the hypothetical outcomes for patients who met rescue criteria on the treated arm. Note that the penalty implicitly accounts for the fact that if such a hypothetical trial was conducted, a portion of data for patients from the treated arm meeting rescue condition $(S = 0)$ and left unrescued would have been missing due to dropouts. That is done through the variance term $\widehat{var}(\hat{\mu}_{(0)})$, which reflects loss of precision due to dropouts in the placebo arm. Therefore, the assumption is that the information loss due to dropout for patients in the treated arm needing rescue is the same as that in the placebo arm.

A similar path is taken in Cro et al. (2019) who use the novel principle of "information-anchored sensitivity analysis" stating that the loss of information due to missing data when conducting sensitivity analysis should be the same as the loss in the primary analysis with missing outcomes (e.g., assuming MAR). For example, the estimator $\widehat{var}(\hat{\delta}_1)$ from Mehrotra, Liu, and Permutt (2017) (as well as equations 22.15 and 22.16) effectively cancels out uncertainty due to missing data by directly imputing the means for missing patterns with those from the reference arm. This results in violating the "information-anchored" principle.

As Cro et al. (2019) argue, a usual imputation strategy for the reference-based sensitivity analysis followed by Rubin's variance estimator does obey this principle. This happens because Rubin's estimator ignores positive correlation between $\hat{\mu}_{(1)|S=0}$ and $\hat{\mu}_{(0)}$ imputed from the same reference arm by treating them as representing two independent study arms and applying standard analyses to each imputed data set. Consequently, Rubin's formula produces variance that is larger than the inherent (true) variance of the imputation-based estimator (which could be evaluated, for example, with bootstrap). However, the conservatism of Rubin's estimator results in obeying the "information-anchored" principle, whereas the correct estimator of the inherent variance of the MI-based point estimator would not.

Each approach has its own assumptions and conceptual considerations. The point here is not to lobby for one approach or another. Rather, the intent is to make clear that practical and conceptual differences exist. It is important to understand the assumptions behind every estimand and associated sensitivity analysis. Therefore, the estimator – and the method for assessing the variance of the estimate – should match the estimand.

22.6 Code Fragments

Code Fragment 22.1 SAS Code for the copy reference method of reference-based imputation

```
PROC MI DATA=y1 SEED=1214 OUT=outmi round=1 NIMPUTE=1000;
        class trt;
    monotone method=reg;
    var basval y1 y2 y3;
        mnar model (y2 / modelobs=(trt='1'));
        mnar model (y3 / modelobs=(trt='1'));
run;
```

```
ods output diffs = diffs;
proc Mixed data=finalY;
    class trt time subject;
    model y= basval trt time trt*time basval*time/
    ddfm=kr outp=check;
    repeated time / subject=subject type = un;
    by _Imputation_;
    LSMEANS trt*time/diffs;
        id subject time ;
run;

data diffs2;
 set diffs;
 if time = _time and _time = 3;
run;

proc mianalyze data=diffs2;
  modeleffects estimate;
  stderr stderr;
run;
```

22.7 Summary

RBI is useful for addressing estimands that involve treatment switching. In reference-based methods, the common idea is to construct a principled set of imputations that model specific changes in treatment. After relevant ICEs, values are imputed by making qualitative reference to another arm in the trial. In other words, imputed post-ICE values have the statistical behavior of otherwise similar subjects in a reference group.

RBI can be implemented using multiple imputation in manners that mimic the immediate or gradual disappearance of a treatment effect after discontinuation of the experimental treatment, as would be applicable to symptomatic treatments. Imputations can also mimic disease modifying treatments by maintaining the treatment effect that was observed at the last visit.

The likelihood- and Bayesian-based approaches to group mean reference-based imputation are also possible. Rather than imputing individual values with uncertainty, as is done in multiple imputation-based approaches, the likelihood- and Bayesian-based methods define the mean of subjects in the experimental group who discontinued to be equal to the mean of the control group. Because imputation is by definition, the variance and standard errors of treatment contrasts reflect no uncertainty in the imputation. Therefore, group mean-based approaches generally yield smaller standard errors than the corresponding MI-based methods. In fact, the standard errors will also be smaller than in the corresponding likelihood-based repeated measures analysis.

Acknowledgment

Sections 22.3 and 22.4 were adapted from the appendix in Mallinckrodt et al. (2019).

23

Delta Adjustment

23.1 Introduction

In any strategy for dealing with intercurrent events (ICEs), it is important to assess sensitivity to assumptions. In strategies where missing or irrelevant data exists because of an ICE, it is important to assess sensitivity to missing data assumptions. As apparent from the definition of missing at random (MAR), the validity of MAR hinges on characteristics of the unobserved data, which of course are unknown. Therefore, assessing robustness of results to departures from MAR remains an important aspect of sensitivity analyses in the ICH E9(R1) framework (Mallinckrodt et al., 2019).

The controlled-imputation family of methods, which includes reference-based imputation (see Chapter 22) and delta adjustment, has gained favor as sensitivity analyses because their assumptions are transparent and easy to understand (O'Kelly and Ratitch, 2014; Mallinckrodt and Lipkovich, 2017). These approaches can also be primary analyses. For example, as described in Chapters 11 and 22, reference-based imputation can be useful as estimators of treatment policies that were not included in the trial or not followed by some subjects. Focus in this chapter is on the use of delta adjustment to assess sensitivity to departures from the conditions assumed in hypothetical strategies for dealing with ICEs.

23.2 Technical Details

Delta adjustment assumes that subjects who discontinued or had irrelevant data because of an ICE had outcomes that were worse than otherwise similar subjects that remained in the study or did not have an ICE. The difference (adjustment) in outcomes between those with and without the relevant ICE can be a shift in mean or slope, an absolute value, or a proportion. The adjustment is referred to as delta (Δ). The delta adjustment can be implemented in manners resulting in increasing, decreasing, or constant departures from MAR (Mallinckrodt and Lipkovich, 2017).

The alteration in the estimated treatment contrast due to delta adjustment or other missing not at random (MNAR) approaches is proportional to the amount of missing data. Therefore, unless missing data rates vary widely between treatments, fitting the same MNAR model to all treatment groups typically has a similar within group effect and therefore a relatively small effect on treatment contrasts. However, different MNAR models for each treatment arm provide greater opportunity for changes to estimated treatment contrasts compared with the corresponding assumed conditions (Mallinckrodt, Lin, and Molenberghs, 2013; Mallinckrodt and Lipkovich, 2017). Delta adjustment is typically implemented by applying delta to only the drug-treated arm(s) of the trial, leaving the control arm unadjusted. However, it is easy to delta adjust each arm, using the same or different deltas (O'Kelly and Ratitch, 2014).

Two general families of delta adjustment exist: marginal and conditional. In the marginal approach, the delta adjustment is applied after imputation of all missing values, and the adjustment at one visit does not influence imputed values at other visits. Therefore, the marginal approach with a constant delta results in a constant departure from assumed conditions. In the conditional approach, the delta adjustment is applied in a sequential, visit-by-visit manner with values imputed as a function of both observed and previously imputed delta-adjusted values. In the conditional approach with delta applied to each visit, the deltas have an accumulating effect over time, resulting in departures from assumed conditions that increase over time. For example, for a subject who drops out after the first post-baseline assessment, the total increment is approximately Delta x (1 + the sum of the correlation between the first post-baseline visit with all subsequent visits). Alternatively, delta applied to only the first missing visit results in departures that decrease over time (Mallinckrodt et al., 2014; O'Kelly and Ratitch, 2014; Mallinckrodt and Lipkovich, 2017).

A single-delta adjustment analysis allows testing whether a specific departure overturns the initial result. However, delta adjustment can be applied repeatedly as a progressive stress test to find how extreme the delta must be to overturn significance of the initial result. This is the so-called tipping point approach (O'Kelly and Ratitch, 2014).

The marginal delta adjustment has a consistent and therefore easy to anticipate effect for a given missing data pattern. The effect of marginal delta adjustment on the endpoint contrast can be analytically determined as follows: Let π = the percentage of missing values at the endpoint visit, and Δ = the delta adjustment applied to the endpoint visit only. The change to the endpoint contrast = $\Delta \times \pi$ (Mallinckrodt and Lipkovich, 2017).

For binary outcomes with imputation based on logistic regression models, the shift parameter Δ is applied to the logistic function, which modifies the predicted probabilities for the classification levels. For an imputed ordinal classification variable, the shift parameter is applied to the cumulative logit function values for the corresponding response level. For an imputed nominal classification variable, the shift parameter is applied to the generalized logit model function values for the corresponding response level (van Buuren, 2007, 2018; Carpenter and Kenward, 2013).

Delta adjustment can also be applied to time-to-event outcomes. The censoring at random (CAR) method of multiple imputation, for example, via Kaplan–Meier imputation (KMI), serves as a starting point for these methods. A single delta adjustment or a tipping point approach can be implemented by adjusting the Kaplan–Meier curve or hazard function by raising it to the power of Δ (Lipkovich, Ratitch, and O'Kelly, 2016).

23.3 Example

To illustrate the delta-adjustment approach, consider the example data set. The SAS code to implement the delta adjustment controlled-imputation approach is listed in Code Fragment 23.1 (Section 23.4). The SAS implementation of delta adjustment uses the conditional approach where delta adjustment at previous imputations influences the current imputation.

The top panel of Table 23.1 summarizes results from applying a delta adjustment of 3 points to the drug-treated arm only, using several approaches. First, adjustment was applied to only the endpoint visit (Time 3). That is,

TABLE 23.1

Results from Various Delta-Adjustment Approaches to the Small Example Data Set with Dropout

	Delta = 3			
Delta	Endpoint Contrast	Standard Error	*p*-Value	Deviation from MAR
No adjustment[a]	2.98	1.71	0.082	0.00
Time 2 only	2.70	1.73	0.119	0.28
Time 3 only	2.25	1.76	0.201	0.73
Time 2 and 3	1.97	1.79	0.271	1.01
Delta Applied to Endpoint Visit Only				
No adjustment[a]	2.98	1.71	0.082	0.00
Delta = 1	2.74	1.73	0.103	0.24
Delta = 2	2.49	1.74	0.152	0.49
Delta = 3	2.25	1.76	0.20	0.73
Delta = 4	2.01	1.79	0.260	0.97
Delta = 5	1.77	1.82	0.323	1.21

[a] Delta = 0, the MAR result. Results do not exactly match previous multiple imputation results due to different seed values.

delta = 0 at Time 2. Even though the SAS implementation uses the conditional method, this implementation yields the same result as a marginal approach. That is, using delta = 0 at Time 2 is equivalent to a Time 2 delta that does not influence Time 3 imputed values. Other implementations applied delta to only Time 2 and applied delta to Time 2 and Time 3. The bottom panel of Table 23.1 summarizes results from assuming progressively larger values of delta applied to only the endpoint visit, which again is equivalent to a marginal approach.

In both panels of Table 23.1, results from using delta = 0, that is no adjustment, are the MAR result to which delta adjustment results are compared. Moreover, interpretation of results is further facilitated by noting the (monotone) missing data patterns of the treatment arm to which the various deltas were applied. In the drug-treated arm (Trt 2), three subjects had missing values at Time 2; those same three subjects plus three additional subjects had missing values at Time 3.

The top panel illustrates that applying the same delta to the endpoint visit has greater impact on the endpoint contrast than applying that delta to missing values at Time 2 only. This stems from two reasons. First, fewer patients have missing data at Time 2 than at Time 3. Hence, delta adjustment is applied to fewer patients. Moreover, applying delta to Time 2 imputed values impacts the endpoint contrast only indirectly via the correlation implied by the imputation model between Time 2 outcomes and Time 3 outcomes.

Applying delta to Time 2 and Time 3 had a greater impact on the endpoint contrast than applying delta to Time 3 only. This is because the total effect of the delta adjustment on the endpoint contrast – in the conditional approach – is essentially the sum of the effect from directly adjusting imputed values for Time 3 and the indirect effect on Time 3 imputed values by having adjusted the Time 2 imputed values that are used as part of the Time 3 imputations.

The bottom panel of Table 23.1 illustrates that increasing the delta applied to the endpoint visit has a consistent and therefore easily predicted effect for a given missing data pattern. In this example, for each 1-point increment of delta, the endpoint contrast was reduced by 0.24 points. This result is intuitive in that 24% of the values were missing for Treatment 2 at Time 3. In this simple illustration of applying what is essentially a marginal delta, the effect of delta adjustment on the endpoint contrast $= \Delta \times \pi$.

From this simple, marginal approach to delta adjustment, it is easy to appreciate several important factors. First, it shows directly the impact of the fraction of missing values on the sensitivity of the MAR result to departures from MAR. If π is cut in half, Δ must double for the adjustment to have the same net effect on the endpoint contrast. It is also easy to see how progressively increasing delta can be used as a stress test to ascertain how severe departures from MAR must be to overturn inferences from the MAR result. However, the progressive stress test is not needed here because in this small example data set, results with dropout were not significant in the MAR result.

23.4 Code Fragments

Code Fragment 23.1 SAS Code for delta adjustment controlled multiple imputation

```
/* 3 point adjustment at last visit only (for treated arm,
   trt=2).  MAR imputation for placebo arm (trt=1).
   No missing values at first postbaseline visit */

PROC MI DATA=y1 (WHERE=(trt=1)) SEED=1214 OUT=outmi_1 ROUND=
1 NIMPUTE=1000;
       MONOTONE METHOD=REG;
       VAR basval y1 y2 y3;
RUN;

PROC MI DATA=y1 (WHERE=(trt=2)) SEED=1214 OUT=outmi_2 ROUND=
1 NIMPUTE=1000;
       CLASS trt;
       MONOTONE METHOD=REG;
       VAR basval y1 y2 y3;
       MNAR ADJUST(y2 /DELTA=0 ADJUSTOBS=(trt='2'));
       MNAR ADJUST(y3 /DELTA=3 ADJUSTOBS=(trt='2'));
RUN;

/* combine imputed data sets data sets outmi_1 and outmi_2
and transpose into a "long" format, data finalY */

ODS OUTPUT diffs = diffs;
PROC MIXED DATA=finalY;
  CLASS trt time subject;
  MODEL y= basval trt time trt*time basval*time/DDFM=kr
  OUTP=check;
  REPEATED time / SUBJECT=subject TYPE = UN;
  BY _Imputation_;
  LSMEANS trt*time/DIFFS;
  ID subject time;
RUN;

DATA diffs2;
 SET diffs;
 IF time = _time AND _time = 3;
RUN;

PROC MIANALYZE DATA=diffs2;
  MODELEFFECTS estimate;
  STDERR stderr;
RUN;
```

23.5 Summary

Delta adjustment is a simple and transparent approach to assess sensitivity to departures from MAR. Delta adjustment assumes that subjects who discontinued or had irrelevant data because of an ICE, had outcomes that were worse than otherwise similar subjects that remained in the study, or did not have an ICE. The adjustment in outcomes between those with and without relevant ICEs can be a shift in mean or slope, an absolute value, or a proportion.

In the marginal approach to delta adjustment, delta is applied after imputation of all missing values and the adjustment at one visit does not influence imputed values at other visits. Therefore, the marginal approach with a constant delta results in a constant departure from MAR (or whatever profile had been assumed, such as jump to reference or copy increment from reference). In the conditional approach, delta is applied in a sequential, visit-by-visit manner with values imputed as a function of both observed and previously imputed delta-adjusted values. In the conditional approach with delta applied to each visit, the deltas have an accumulating effect over time, resulting in departures from assumed conditions that increase over time.

A single delta adjustment analysis allows testing whether a specific departure overturns the initial result. In the tipping point approach, delta is sequentially increased as a progressive stress test to find how extreme the delta must be to overturn significance of the initial result.

24

Overview of Principal Stratification Methods

24.1 Introduction

Foundational references for principal stratification include Frangakis and Rubin (2002); Zhang and Rubin (2003); Barnard et al. (2003); and Roy, Hogan, and Marcus (2008).

As noted in Chapter 6, in parallel group designs, only one potential outcome can be observed per subject. However, in many instances, it would be desirable to observe multiple potential outcomes; for example, the outcome of patient i on both the experimental and control treatments as would be possible in a cross-over design. Randomization facilitates inferences about the difference between experimental and control treatments when each subject receives only one treatment. However, intercurrent events (ICEs) can effectively break the randomization, thereby jeopardizing the validity of causal inferences about the treatments. Principal stratification strategies can be useful for these situations although additional identifying assumptions are needed for estimation. In this chapter, several scenarios relevant to clinical trials are considered.

Distinction is drawn between post-randomization outcomes S (e.g., adverse events) and post-randomization changes in treatment, R (e.g., changes in dose, addition of or switch to rescue medication). Consideration is given to the principal strata based on $R(1), R(0)$ and on the principal strata based on $S(1), S(0)$. Estimates of treatment effect within groups based on principal strata are causal because $S(1), S(0)$ and $R(1), R(0)$ are (by definition) not affected by treatment assignment (Z) and therefore can be modeled similar to pretreatment covariates.

24.2 Principal Stratification Based on Intercurrent Events

The principal stratification strategy can account for ICEs by assessing treatment effects within the relevant stratum, such as in patients who would be compliant to or remain alive on all treatments in the study. An example approach implementing this analysis would be to first estimate whether each patient would have a relevant ICE(s) using a logistic regression model with baseline demographic and prognostic factors as independent variables. Patients are assigned to strata based on the predicted probabilities. Data from the relevant stratum is then analyzed and treatment groups are compared based on the population parameters using an estimation method and model appropriate for the endpoint and design. Principal stratification can be applied to continuous, categorical, and time-to-event endpoints.

A method suggested by Rubin (1998) for the ICE of death is applicable to the general scenario noted in the previous paragraph. This approach can be described using the terminology of potential outcomes and principal stratification. For the ith patient, denote by $D_i(z) = 0, 1$ the survival outcome (alive or dead, respectively) at the end of planned follow-up on treatment $z = 0, 1$, by $Y_i(z)$ = the score on treatment z targeted by the estimand, and by X = a set of baseline covariates. In the following, we drop the patient's index unless it may cause confusion.

Each patient has two sets of potential outcomes:

- $(D(0), Y(0))$ – survival and score on control treatment
- $(D(1), Y(1))$ – survival and score on experimental treatment

Four principal strata can be formed with respect to survival status:

1. Patients who would survive to the end of planned follow-up on either treatment: $LL = \{i | D(0) = D(1) = 0\}$
2. Patients who would die before the end of planned follow-up on either treatment: $DD = \{i | D(0) = D(1) = 1\}$
3. Patients who would survive until the end of planned follow-up on control but would die on experimental treatment: $LD = \{i | D(0) = 0, D(1) = 1\}$
4. Patients who would survive until the end of planned follow-up on experimental but would die on control treatment: $DL = \{i | D(0) = 1, D(1) = 0\}$

The principal stratum LL is the only group for which a causal estimand for Y involves well-defined values of Y, whereas in other strata, Y does not have a meaningful value under one or both treatments. An estimate of a causal effect of treatment on Y in the LL stratum is termed the survivor average causal effect (SACE).

A practical approach to estimating SACE in a randomized study is described in Rubin (1998, Section 6). The analysis would proceed as follows, using multiple imputation methodology:

- Each patient has an observed survival status at the end of planned follow-up for treatment z when it is the same as the treatment to which they were randomized, z_i, denoted by $D_i(z_i)$. This involves the so-called consistency assumption that observed outcomes are equal to potential outcomes associated with the treatment that was actually assigned to a given patient.

  Patients' survival status under the opposite treatment is not observed: $D_i(1-z_i)$.

- Estimate a multiple imputation model within each treatment group separately based on the baseline covariates X and observed survival status $D_i(z_i)$. A logistic regression model for the binary survival status can be used. An alternative that uses more information could impute time to death using, for example, a Cox regression or a piecewise exponential survival model (Lipkovich, Ratitch, and O'Kelly, 2016).

- For each patient on control treatment with $z_i = 0$, predict (impute) survival status on the experimental treatment, $\hat{D}_i(1)$ using the multiple imputation model estimated from the experimental group, given patients' baseline covariates x_i.

- For each patient on the experimental treatment $(z_i = 1)$, predict (impute) survival status on the control treatment, $\hat{D}_i(0)$ using the multiple imputation model estimated from the control group, given patients' baseline covariates x_j.

  Create multiple (M) imputed data sets.

- In each $m = 1, \ldots, M$ imputed data set, select patients who are observed/predicted to survive on both treatments, i.e.:

$$LL^m = \left\{ i \left| \begin{array}{l} D_i(0) = 0 \text{ and } \hat{D}_i(1) = 0, \text{ if } z_i = 0 \text{ or} \\ D_j(1) = 0 \text{ and } \hat{D}_i(0) = 0, \text{ if } z_i = 1 \end{array} \right. \right\}.$$

Using the set of patients LL^m, estimate the difference between treatments in change from baseline at the end of planned follow-up, for example, using an analysis of covariance (ANCOVA) analysis.

- Combine treatment difference estimates from the M data sets using Rubin's combination rules.

Survival probabilities (or time to death) must be predictable by baseline covariates (explainable nonrandom survival), and an unverifiable assumption of no unobserved confounders must be made. Also, if there are very few deaths at the end of planned follow-up, it might be difficult to estimate a robust model. Rubin's suggested approach described above could be extended for use with other ICEs.

24.3 Principal Stratification Based on Post-randomization Treatment

One of the first applications of principal stratification was the complier average causal effect (CACE; see, e.g., Imbens and Rubin (1997)). Here, interest is in the treatment effect within the stratum of patients who would be compliant with either (all) treatment. To illustrate, consider a 2 by 2 table based on the cross-tabulation of potential outcomes $R(1), R(0)$ for the post-randomization treatment. The shaded cell is excluded by the assumption of monotonicity (that is explained in the following).

	$R(1) = 0$	$R(1) = 1$
$R(0) = 0$	Never takers (N)	Compliers (C)
$R(0) = 1$	Defiers (D)	Always takers (A)

The cells can be described as:

- *Compliant with experimental but not control* "Compliers," $R(0) = 0$, $R(1) = 1$
- *Compliant for both treatments* "Always takers," $R(1) = 1, R(0) = 1$
- *Not compliant for both treatments* "Never takers," $R(0) = 0, R(1) = 0$
- *Compliant with control but not experimental* "Defiers," $R(0) = 1, R(1) = 0$

Let the true proportions of patients in the four cells of the cross-tabulation be $p_{00}, p_{01}, p_{10}, p_{11}$. Although the probabilities for each cell are generally unknown, the marginal probabilities are estimable from the observed data. For example, $P(R(0) = 0) = p_{00} + p_{01}$ can be estimated unbiasedly as the proportion of patients randomized to placebo who remained untreated at the end of the study.

Assume interest is in estimating the expected difference in potential outcomes for $Y(1)$ and $Y(0)$ in compliers only. That is, $E(Y(1) - Y(0) | R(0) = 0, R(1) = 1)$ or using letter codes $E(Y(1) - Y(0) | C)$. This can be accomplished using observed outcome Y after assuming the usual stable unit treatment value assumption (SUTVA) and assuming patients are randomly assigned to treatment arms, so all potential outcomes are independent of $Z = 0, 1$. Moreover, the following assumptions are made:

1. "Exclusion restriction," requiring for never takers and always takers that potential outcomes are the same, regardless of what arm they were randomized to, that is $Y(1) = Y(0)$.

2. "Monotonicity," $R(1) \geq R(0)$, which simply means that there are no "defiers" (i.e., the combination $R(0) = 1$ and $R(1) = 0$ is excluded and $p_{10} = 0$).

3. Positivity, the probability of membership in the stratum of interest (here, compliers) is positive, $p_{01} > 0$.

Because there are no "defiers" (Assumption 2) and the treatment difference in never takers and always takers is 0, the overall treatment effect is as a weighted average of expected outcomes across the four strata (C, N, A, D).

$$E\big(Y(1) - Y(0)\big) = E\big(Y(1) - Y(0)|\,C\big) \times p_{01}$$

$$+ E\big(Y(1) - Y(0)|\,N\big) \times p_{00} + E\big(Y(1) - Y(0)|\,A\big) \times p_{11} + E\big(Y(1) - Y(0)|\,D\big) \times p_{10}$$

$$= E\big(Y(1) - Y(0)|\,C\big) \times p_{01} + 0 \times p_{00} + 0 \times p_{11} + E\big(Y(1) - Y(0)|\,D\big) \times 0$$

$$E\big(Y(1) - Y(0)|\,C\big) = \frac{E\big(Y(1) - Y(0)\big)}{p_{01}}$$

The numerator can be estimated via the usual intent-to-treat estimator (by the randomization and SUTVA assumptions). The denominator is identifiable via assumption 2. Because the probability p_{10} is 0, and the marginal probabilities are estimable, and p_{00}, p_{01}, p_{11} can be estimated by subtraction.

Estimation is simplified when the experimental treatment is compared to placebo rather than another active treatment. Here, the outcome for patients who were randomized to the experimental treatment and did not comply (e.g., discontinued from treatment/study) can be assumed to be similar to patients randomized to placebo (Assumption 1).

24.4 Principal Stratification Based on the Post-randomization Outcomes

24.4.1 Introduction

Assume the post-randomization event is a symptom S and that the goal is to estimate the treatment effect in the stratum defined by $S(1), S(0)$. A special case is when outcomes can be measured only in patients who experienced the post-randomization event. For example, in a vaccine trial interest may be in assessing viral load in patients who became infected. At the time of randomization to the vaccine or control, it is not known who will become infected and probability of infection is unlikely to be independent of treatment.

24.4.2 The Case of a Binary Outcome

The principal strata are illustrated in the cross-tabulation below, which is similar to the previous cross-tabulation except here the post-randomization event is presence or absence of a harmful symptom rather than a type of treatment taken; hence the descriptive names for the four cells are different. The shaded cell is excluded by the assumption of monotonicity.

	$S(1)=0$	$S(1)=1$
$S(0)=0$	Immune (I)	Harmed (H)
$S(0)=1$	Benefiters (B)	Doomed (D)

- "Doomed," $S(1)=1, S(0)=1$
- "Benefiters," $S(0)=1, S(1)=0$
- "Harmed," $S(0)=0, S(1)=1$
- "Immune," $S(0)=0, S(1)=0$

The stratum of interest is the treatment effect for the patients in the upper left cell, "Immune."

The identifiability assumptions are similar to those in Section 24.3, except there is no analogue to Assumption 1, the exclusion restriction.

1. "Monotonicity," $S(1) \leq S(0)$; in the present context, this means that everyone who had the symptom on treatment would have also had it if randomized to placebo; therefore, there are no "harmed" by treatment, and the combination $S(0)=0$ and $S(1)=1$ is excluded.

2. A positive probability in the stratum of interest (Immune) is non-zero, $p_{00} > 0$.

To make connection with the case when the undesirable symptom is an adverse event, the substratum of interest would still be "Immune," whereas the monotonicity assumption would be reversed, $S(1) \geq S(0)$, implying that everyone who had the symptom (adverse event) on placebo would have had it if treated, making "Benefiters" the empty cell.

The goal is to estimate the odds ratio in the "Immune" stratum.

$$OR(I) = \frac{P(Y(1)=1|I)(1-P(Y(0)=1|I))}{(1-P(Y(1)=1|I))P(Y(0)=1|I)}$$

The null hypothesis of interest is $OR(I) \geq 1$ against the alternative $OR(I) < 1$ (assuming the desirable outcome is $Y = 0$).

Because the stratum of "Harmed" by treatment is assumed empty the conditional probability of potential outcomes $Y(0)$ in the stratum $I = (S(0) = 0)$ is estimable using observed data (within the subset $Z = 0, S = 0$). However, it is only possible to estimate the probability of $Y(1)$ in a combined substratum I or B, $P(Y(1) = 1| I \text{ or } B)$.

Following the example in Magnusson et al. (2018), one strategy may be to express $P(Y(1) = 1| I)$ via a sensitivity parameter

$$\gamma = P(Y(1) = 1| B)$$

and use the latter in the sensitivity analysis of the $OR(I)$.

Since,

$$P(Y(1) = 1| I \text{ or } B) = P(Y(1) = 1| I) P(I| I \text{ or } B) + P(Y(1) = 1| B) P(B| I \text{ or } B)$$

The probability of binary outcome in "Immune," $P(Y(1) = 1| I)$ can be expressed as

$$P(Y(1) = 1| I) = \frac{P(Y(1) = 1| I \text{ or } B)}{P(I| I \text{ or } B)} - \frac{P(B| I \text{ or } B)}{P(I| I \text{ or } B)} \times \gamma,$$

All quantities except the sensitivity parameter, γ, can be estimated from the observed data; therefore the estimated $\widehat{OR}_\gamma(I)$ can be written as a function of γ and stress testing can be done by varying γ within a plausible range or from 0 to 1. If the size of set I versus B is very large, the range of resulting values for $P(Y(1) = 1| I)$ will be narrow, for any assumed γ. Magnusson et al. (2018) further proposed a Bayesian framework for conducting sensitivity analysis, where the identifiably is ensured by specifying strongly informative prior distributions on specific parameters.

An alternative specification of the sensitivity parameter suggested by Mehrotra, Li, and Gilbert (2006) is:

$$P(Y(1) = 1| I) = \frac{P(Y(1) = 1| I \text{ or } B) P(I| Y(1) = 1, I \text{ or } B)}{P(I| I \text{ or } B)}$$

The only unidentifiable quantity is $P(I| Y(1) = 1, I \text{ or } B)$, which can be determined via a sensitivity parameter β with a logistic specification

$$P(I| Y(1) = y, I \text{ or } B) = \frac{1}{1 + \exp(-\alpha - \beta y)},$$

where $y \in \{0,1\}$.

With β fixed at any value, the parameter α can be identified from the constraint:

$$\frac{P(I\,\text{or}\,B)}{P(I)}\sum_{y=0}^{1}P\big(I|\,Y(1)=y,\,I\,\text{or}\,B\big)P\big(Y(1)=y|\,I\,\text{or}\,B\big)=1.$$

Perhaps specifying a sensitivity parameter β is more convenient, and it has a more natural interpretation as a log odds ratio. Essentially it captures the belief in how a treated patient's positive/negative outcome influences the odds that the patient would be immune, regardless of his or her treatment assignment. For example, if $\beta < 0$, then those with better outcomes in the treated arm $(Y = 0)$ are more likely to be those who would have been immune if randomized to placebo.

24.4.3 The Case of a Continuous Outcome

Consider the case of continuous outcomes using the framework of Mehrotra, Li, and Gilbert (2006). Conceptually, the cross-tabulations to identify strata apply as before, except the outcome variable Y, is continuous (with lower values clinically desirable). As before, interest is in evaluating the treatment effect within the "Immune" stratum, $I = (S(0) = 0\ \&S(1) = 0) \equiv (S(0) = 0)$, by the monotonicity assumption.

In its most general form, the null hypothesis is formulated in terms of comparing distributions of potential outcomes (within the "Immune" stratum):

$$F_0\big(y\,|\,I\big)=\Pr\big(Y(0)\le y\,|\,I\big)\ \text{vs.}\ F_1\big(y\,|\,I\big)=\Pr\big(Y(1)\le y\,|\,I\big),\ \text{for all }y.$$

As before, $F_0(y\,|\,I)$ can be estimated from the observed data (in the placebo arm for patients with no symptom, $Z = 0, S = 0$), whereas $F_1(y\,|\,I)$ cannot be estimated and requires a sensitivity parameter that (as in the binomial case) can be expressed within the conditional distribution of $S(0)$ given $Y(1)$.

Specifically, using probability calculus, we can write

$$F_1\big(y\,|\,I\big)=\int_{-\infty}^{y}f_{Y(1)}\big(t|\,S(1)=0,S(0)=0\big)dt$$

$$=\int_{-\infty}^{y}f_{Y(1)}\big(t|\,S(1)=0\big)\frac{P\big(S(0)=0|\,S(1)=0,Y(1)=t\big)}{P\big(S(0)=0|\,S(1)=0\big)}dt$$

$$=\frac{P\big(S(1)=0\big)}{P\big(S(0)=0\big)}\int_{-\infty}^{y}f_{Y(1)}\big(t|\,S(1)=0\big)w(t)dt,$$

where the weight $w(\cdot)$ is a function of potential outcome $Y(1)$, which can be assumed to have a logistic form as

$$w(y) = P\big(S(0) = 0 \mid S(1) = 0, Y(1) = y\big) = \frac{1}{1 + \exp\big(-\alpha - \beta y\big)}$$

The conditional density $f_{Y(1)}(t \mid S(1) = 0)$ can be estimated from the observed data (parametrically or nonparametrically) using Y for patients randomized to the treatment arm ($Z = 1$) who had no symptom ($S = 0$). Similarly, $P(S(1) = 0)$ and $P(S(0) = 0)$ are estimated as the proportions of patients without the symptom in treated and control arms, respectively.

The intercept α in the logistic model can be identified for any given β from the constraint that the full integral for $f_1(y \mid I)$ must be unity,

$$\frac{P\big(S(1) = 0\big)}{P\big(S(0) = 0\big)} \int_{-\infty}^{\infty} f_{Y(1)}\big(t \mid S(1) = 0\big) w(t) \, dt = 1$$

The slope parameter, β, is a sensitivity parameter that quantifies the log odds for not having the symptom if randomized to the placebo arm, per one unit increase in outcome when on treatment (and given no symptom on treatment). The following statistic for testing the hypotheses of the equality of distributions $F_0(y \mid I)$ and $F_1(y \mid I)$ based on the above formulation can be adopted from the proposal in Lu, Mehrotra, and Shepherd (2013):

$$T(\beta) = \int_{-\infty}^{\infty} y \big\{ d\hat{F}_0(y \mid I) - d\hat{F}_1(y; \beta \mid I) \big\}$$

$$= \frac{1}{n_0} \sum_{i=1}^{N} (1 - z_i)(1 - s_i) Y_i - \frac{\hat{p}_1}{\hat{p}_0} \frac{1}{n_1} \sum_{i=1}^{N} z_i (1 - s_i) w\big(y_i; \hat{\alpha}, \beta\big) y_i,$$

where N is the total number of patients in the study, n_0 and n_1 are the number of patients with $S = 0$ in placebo and treated groups, respectively; $\hat{p}_0$ and $\hat{p}_1$ are the proportion of patients without symptom S in the control and treated arms, respectively.

For any specified value of β, $T(\beta)$ can be considered a bias-corrected version of a test statistic comparing group means in patients who had no symptom, where treated patients who have higher probability of no symptom if randomized to placebo are upweighted, and patients with smaller probability are downweighted. The standard errors can be computed using bootstrap methods. Lu, Mehrotra, and Shepherd (2013) also proposed a rank-based version of $T(\beta)$.

Shepherd, Gilbert, and Dupont (2011) proposed a framework for sensitivity analysis extending ideas of Mehrotra, Li, and Gilbert (2006) by relaxing the monotonicity assumption. This results in an additional sensitivity parameter that captures the degrees of departure from monotonicity. They also consider a time-to-event outcome that brings additional challenges in that membership in stratum $S(z)$ can be unknown even for patients randomized to $Z = z$ because of censoring. They suggested estimating the probability of strata membership via a nonparametric Kaplan–Meier estimator assuming censoring is noninformative.

An alternative way to introducing sensitivity parameters in the context of principal stratification is in obtaining bounds for the unknown parameters of interest, given observed data. Since unobservable quantities can be expressed as mixtures of latent outcomes across principal strata, one can easily derive bounding inequalities for unobservables in terms of observable quantities. These boundaries can be tightened by utilizing available baseline covariates (under some assumptions). For details and examples, see Miratrix et al. (2018) and references therein.

24.5 Utilizing Baseline Covariates

24.5.1 Introduction

One way to estimate outcomes in principal strata is by employing information contained in baseline covariates X; then assume (rather strong assumption) that given X, the membership in $S(0)$ (missing in the treated arm) would provide no additional information for predicting potential outcome $Y(1)$. Symbolically, $S(0) \perp Y(1) \mid X$. Similarly, knowing $S(1)$ (missing in the placebo arm) provides no additional information for predicting potential outcome $Y(0)$, after X has been fitted.

This assumption is known as "principal ignorability" (see Jo and Stuart, 2009; Ding and Lu, 2016) and can be compared with "ignorable treatment assignment" (or "no unmeasured confounder") assumption in observational trials that potential outcomes $Y(0), Y(1)$ are independent of actual (nonrandom) treatment assignment Z, given pretreatment covariates X. Analogously, principal ignorability assumes that potential outcomes are independent of principal strata membership, given covariates.

The main goal is again to identify the distribution of outcomes in the "immune" stratum, which is under the monotonicity assumption:

$$F_1(y \mid I) = \int_{-\infty}^{y} f_{Y(1)}\big(t \mid S(0) = 0\big) dt$$

Following Bornkamp and Bermann (2019), we can rewrite the above in two different forms that give rise to two different approaches for estimating the treatment effect in strata: based on (i) predicted counterfactual response ("placebo response" for treated in the setting of Bornkamp and Bermann (2019)) or (ii) weighting by propensity of strata in the placebo arm.

24.5.2 Predicted Counterfactual Response

Assuming conditional independence of $Y(1)$ and $S(0)$, given X,

$$F_1(y \mid I) = \int_{-\infty}^{y} f_{Y(1)}(t \mid S(0) = 0) dt$$

$$= \int_{-\infty}^{y} \int_{x} f_{Y(1)}(t \mid S(0) = 0, X = x) f(x \mid S(0) = 0) dx dt$$

$$= \int_{-\infty}^{y} \int_{x} f_{Y(1)}(t \mid X = x) f(x \mid S(0) = 0) dx dt$$

$$= \int_{-\infty}^{y} \int_{x} f_Y(t \mid X = x, Z = 1) f(x \mid S = 0, Z = 0) dx dt$$

Note that in the last line, we used independence of potential outcomes for $Y(z)$ in actual treatment assignment Z as well as the consistency assumption that under treatment $Z = z$, the observed outcome Y is the same as potential outcome $Y(z)$.

Using a general form of the test statistic (see Section 24.4.3)

$$T = \int_{-\infty}^{\infty} y \{ d\hat{F}_0(y \mid I) - d\hat{F}_1(y; \beta \mid I) \},$$

Observe that the first component of the difference can be estimated as the simple average of the outcome for patients randomized to the placebo arm ($Z = 0$) who had no symptom ($S = 0$). The second component can be evaluated using predicted responses for patients in the same subset $\{Z = 0, S = 0\}$ if they were treated (contrary to the fact).

This suggests the following test statistic:

$$T_1 = \frac{1}{n_0} \sum_{i=1}^{N} (1 - z_i)(1 - s_i) \left(y_i - \hat{m}_1(x_i) \right),$$

where $\hat{m}_1(x_i)$ is "predicted counterfactual response" based on a regression of Y on X estimated from all treated patients, which is evaluated at covariate profile x_i for each placebo patient in group $S = 0$. The regression $\hat{m}_1$ can be a simple linear regression or constructed using sophisticated machine learning methods.

24.5.3 Strata Propensity Weighted Estimator

The expression for $F_1(y \mid I)$ in the previous section can be re-written as

$$F_1(y \mid I) = \int_{-\infty}^{y} \int_x f_Y(t \mid X = x, Z = 1) f(x \mid S = 0, Z = 0) dx dt$$

$$= \int_{-\infty}^{y} \int_x f_Y(t \mid X = x, Z = 1) P(S = 0 \mid Z = 0, X = x) \frac{f(x \mid Z = 0)}{P(S = 0 \mid Z = 0)} dx dt$$

$$= \int_{-\infty}^{y} \int_x f_Y(t \mid X = x, Z = 1) w_0(x) \frac{f(x \mid Z = 0)}{P(S = 0 \mid Z = 0)} dx dt,$$

where $w_z(x)$ is the probability of observing $S = 0$, given covariates for treatment $Z = z$, which can be estimated using logistic regression. We note that $w_z(x)$ is closely related to principal scores, which are similar to propensity scores for estimating treatment effects in observational studies with nonrandom treatment assignment (see Jo and Stuart, 2009 and Ding and Lu, 2016). Like propensity scores, the principal score is a balancing score in that the distribution of covariates is balanced across principal strata (here $S(0) = 1$ and $S(0) = 0$) conditional on the principal score.

This suggests the following test statistic:

$$T_2 = \frac{1}{n_0} \sum_{i=1}^{N} (1 - z_i)(1 - s_i) y_i - \frac{1}{N_0} \sum_{i=1}^{N} (1 - z_i) \frac{1}{\hat{P}(S = 0 \mid Z = 0)} \hat{w}_0(x_i) \hat{m}_1(x_i),$$

where N_0 is the number of placebo patients, $\hat{w}_0(x_i)$ is the estimated probability of $S = 0$, based on a logistic regression fitted to the placebo arm, and evaluated on placebo patient with covariate profile x_i. And $\hat{m}_1(x_i)$ is as before.

24.6 Implementing Principal Stratification Strategy via Imputation

An attractive strategy for evaluating various estimands based on principal stratification is multiple imputation. Since principal strata (as any potential outcomes) can be formulated as a missing data problem, various multiple imputation (MI) strategies can be implemented. Here missing data is formed by an unknown post-randomization event $S(1)$, for $Z = 0$ and $S(0)$, for $Z = 1$. Again, assume that baseline covariates contain useful information for estimating principal strata.

One general strategy is given as follows:

1. Fit a Bayesian logistic regression to observed data (S, Z, X), where S is the column-vector for symptoms, $Z = \{0,1\}$ is the data column with treatment assignment at randomization, and X is N by p-dimensional matrix of baseline covariates. We will denote X_0 for an N_0 by p covariate matrix for placebo and X_1 the N_1 by p matrix for treated $(N_0 + N_1 = N)$.

$$Logit[P(S = 1 | Z = z, X = x) = a_0 + a_1 z + a_2^T x$$

2. Make M draws from a posterior distribution of parameters $\tilde{a}_0, \tilde{a}_1, \tilde{a}_2$; for each draw, m

   a. Compute the posterior probability of $S = 1$, given "fake covariates" formed as $(Z = 1, X_0)$ and $(Z = 0, X_1)$, that is, for patients from untreated and treated arms with covariate vectors x_{0i} and x_{1j}, the posterior probabilities are $\tilde{p}(1, x_{0i})$ and $\tilde{p}(0, x_{1j})$, respectively.

   b. Impute missing $S(0), S(1)$ by generating Bernoulli random variates from $\tilde{p}(1, x_{0i}), i = 1, ..., N_0$, and $\tilde{p}(0, x_{1i}), i = 1, ..., N_1$.

   c. Based on observed and imputed $S(0), S(1)$, form a subset based on desired principal strata, for example, $I = (S(0) = 0 \ \& \ S(1) = 0)$

   d. Estimate treatment effect within subset I from the current draw $\hat{\delta}_m(I)$

3. Compute MI point estimates $\hat{\delta}(I) = M^{-1} \sum_{m=1}^{M} \hat{\delta}_m(I)$

4. Standard errors can be computed by bootstrap or Rubin's rule.

Although this imputation strategy does not require nor assume the monotonicity constraint, it can be incorporated into the imputation model. For example, estimate the imputation model under the constraint that the $logit[P(S | Z = 1, x)] \leq logit[P(S | Z = 0, x)]$, which implies $a_1 \leq 0$. This assumption can be tested from observed data, assuming that the model including covariates is correctly specified – which would rarely be the case.

To estimate treatment effects in a subset I, another imputation model may be needed to impute possibly missing outcomes Y. Imputed/observed $S(0), S(1)$ can be used as covariates in the imputation model, in case imputations are done across all principal strata (if outcomes of interest are only within a specific stratum, they can be selected postimputation).

Post-baseline outcomes or Y cannot be used in modeling strata membership because including other post-baseline outcomes V for a treated patient when predicting $S(0)$ would also require including $V(0)$ as a predictor, which is unknown. Similarly to the sensitivity parameters in previous sections, a sensitivity parameter β can be added to the imputation model, when generating $S(0)$ for *treated* patients with no symptom $(S(1) = 0)$.

$$\tilde{p}(0, x_{1i}) = \tilde{a}_0 + \tilde{a}_2^T x_{1i} + \beta y_i I(s_i = 0),$$

$$S_i(0) \sim Ber\left(\tilde{p}(0, x_{1i})\right),$$

where y_i is observed or imputed value of the outcome and β is the sensitivity parameter.

24.7 Approximation of Survivor Average Causal Effect

The survivor average causal effect (SACE) has been discussed (Zhang and Rubin, 2003) in the context of estimating a causal effect in the presence of post-randomization deaths. Denoting by $S(z) = 1$ the status that a patient survives to a time point of interest under treatment z, there are four principal strata, as described below.

Principal strata defined by post-randomization survival status.

	$S(1) = 0$	$S(1) = 1$	
$S(0) = 0$	Doomed	Experimental-only survivors	
$S(0) = 1$	Control-only survivors	Always survivors	All control survivors
		All experimental survivors	

Interest is in the treatment effect in the principal stratum of patients who would survive to a time point of interest on either treatment (always survivors).

$$SACE = E[Y(1) - Y(0) \,|\, S(1) = S(0) = 1],$$

where $S(z) = 1$ represents survival to some time point K under treatment z.

A crude estimate or approximation of the SACE has been proposed (Chiba and VanderWeele, 2011). This estimator is the treatment effect estimated from the observed survivors minus a sensitivity parameter:

$$SACE = E\left[Y(1) | Z = 1, S = 1\right] - E\left[Y(0) | Z = 0, S = 1\right] - \alpha,$$

where $\alpha = E[Y(1)|Z=1, S=1] - E[Y(1)|Z=0, S=1]$. The sensitivity parameter α represents the average difference in the outcome that would have been observed under the experimental treatment, $Z=1$, comparing two populations: the first is the population that would have survived with the experimental treatment ($Z=1, S=1$) and the second is the population that would have survived without the experimental treatment ($Z=0, S=1$). This estimate would be a conservative estimate of SACE under the following assumptions:

- Monotonicity: $S(0) \leq S(1)$ for all patients, that is, survival under the experimental treatment is at least as good as under the control treatment, and there is no heterogeneity of the treatment effect on survival in the population. This renders the stratum in the shaded cell of "control-only survivors" empty.

- $\alpha \leq 0$, that is, that the subset of survivors under the control treatment would have better outcomes on the experimental treatment than the population of survivors under the experimental treatment. In other words, it assumes that the control treatment survivors are healthier overall than experimental treatment survivors and that the experimental treatment would never worsen their outcomes $Y(1)$.

The sensitivity parameter α would need to be specified based on expert opinion and is not estimated from data.

The approach described in Colantuoni et al. (2018) provides lower and upper bounds on a crude estimate of SACE under certain assumptions:

$$Lower_{SACE} \leq E\left[Y(1)|S(1) = S(0) = 1\right] - E[Y(0)|S(1) = S(0) = 1] \leq Upper_{SACE}$$

The average outcome in the control group, $E[Y(0)|S(1) = S(0) = 1]$, can be estimated from the observed survivors under the control treatment based on the monotonicity assumption, that is, that the control treatment survivors are also experimental treatment survivors.

The average outcome in the experimental group, $E[Y(1)|S(1) = S(0) = 1]$, can be bounded from below and above as follows. The observed experimental survivors are a mix of "always survivors" and "experimental-only survivors." Assuming the "always survivors" have better outcomes than the "experimental-only survivors":

- The lower bound can be estimated from all observed experimental survivors, $E[Y(1)|A = 1, S(1) = 1]$. Essentially, it means that the average observed outcome in all actual experimental survivors is an estimate of average $Y(1)$ in "always survivors" diluted by presumably nonbetter outcomes of "experimental-only survivors."

- The upper bound can be estimated from $n \times q$ best values of outcome Y observed in the experimental group, where the proportion q is the proportion of always survivors, which can be estimated as the proportion of observed survivors in the control group (p_{00} equals to the number of observed control survivors divided by the number of randomized to the control group).

The upper bound estimate above is reminiscent of a trimmed mean approach (Permutt and Li, 2017).

24.8 Summary

The language of potential outcomes that is widely accepted in the causal inference literature is uncommon within the clinical trialist community and was not used in defining causal estimands in the National Research Council (NRC) report or the ICH E9(R1). However, bridging the gap between the causal inference community and clinical trialists can further advance the use of causal estimands in clinical trial settings. Concepts from causal literature, such as potential outcomes and principal stratification, can greatly facilitate defining and implementing causal estimands and may provide a unifying language to describing the targets for both observational and randomized clinical trials.

Section V

Case Studies
Detailed Analytic Examples

Section V presents case studies from diverse settings that illustrate many of the concepts presented throughout this book.

25

Analytic Case Study of Depression Clinical Trials

25.1 Introduction

The aim of this case study is to compare an experimental drug versus placebo in treating patients with major depressive disorder (MDD) based on mean change from baseline to endpoint in all randomized patients, using the hypothetical strategy of dealing with the intercurrent event of early discontinuation by estimating what the mean changes would have been if all patients had remained adherent. Sensitivity analyses to assess robustness of conclusions to the assumptions in this analysis are also illustrated. This case study is taken from Mallinckrodt and Lipkovich (2017).

A secondary goal of this case study is to illustrate another hypothetical strategy for dealing with early discontinuation in which a reference-based imputation approach is used to estimate what would have happened if patients who discontinued early had remained in the study on placebo treatment. The intent of this analysis is to assess the treatment effect assuming that patients who discontinued drug had zero benefit, which is accomplished by assigning the patients who discontinue early placebo-like outcomes using reference-based imputation.

This case study involves two data sets that were based on actual clinical trial data but were somewhat contrived to avoid implications for marketed drugs tested in those studies. Nevertheless, the key features of the original data were preserved. The original data were from two nearly identically designed clinical trials in patients with MDD that were originally reported by Goldstein et al. (2004) and Detke et al. (2004). Each trial had four treatment arms with approximately 90 patients per arm. The treatment arms included two doses of an experimental medication (subsequently granted marketing authorizations in most major jurisdictions), an approved medication, and placebo. Assessments on the HAMD17 (Hamilton 17-item rating scale for depression) (Hamilton, 1960) were taken at baseline and weeks 1, 2, 4, 6, and 8 in each trial.

All patients from the original placebo arm were included along with a contrived drug arm that was created by randomly selecting 100 patients from the nonplacebo arms. In addition to including all the original placebo-treated patients, additional placebo-treated patients were randomly re-selected so that there were also 100 patients in the contrived placebo arms. A new identification number was assigned to the re-selected placebo-treated patients, and outcomes were adjusted to create new observations by adding a randomly generated value to each original observation.

These trials are referred to as the low- and high-dropout MDD data sets. In the high-dropout data set, completion rates were 70% for drug and 60% for placebo (see Table 25.1). In the low-dropout data set, completion rates were 92% in both the drug and placebo arms. The dropout rates in the contrived data sets closely mirrored those in the corresponding original studies. The design differences that may explain the difference in dropout rates between these two otherwise similar trials were that the low-dropout data set came from a study conducted in Eastern Europe that included a 6-month extension treatment period after the 8-week acute treatment phase and used titration dosing. The high-dropout data set came from a study conducted in the United States that did not have the extension treatment period and used fixed dosing.

Visitwise mean changes for patients that completed the trials versus those who discontinued early are summarized in Figures 25.1 and 25.2 for the low- and high-dropout data sets, respectively. In the high-dropout data set, patients who discontinued early had less favorable outcomes than completers. With only a few dropouts at each visit in the low-dropout data set, trends were not readily identifiable.

To assess the primary estimand, mean changes from baseline were analyzed using a restricted maximum likelihood (REML)-based repeated measures approach. The analysis included the fixed, categorical effects of treatment, investigative site (site), visit, treatment-by-visit interaction, and site-by-visit interaction, along with the continuous, fixed covariates of baseline score and

TABLE 25.1

Number of Observations by Week in MDD Data Sets

Week	High Dropout					Low Dropout				
	1	2	4	6	8	1	2	4	6	8
Placebo	100	92	85	73	60	100	98	98	95	92
Drug	100	91	85	75	70	100	98	95	93	92

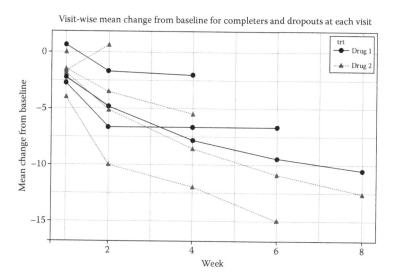

FIGURE 25.1
Visitwise mean changes from baseline by treatment group and time of last observation in the low-dropout data set.

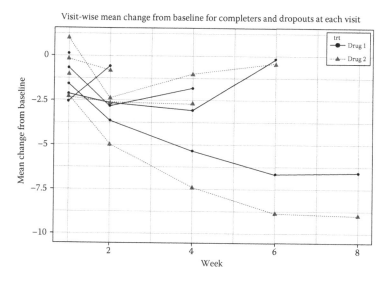

FIGURE 25.2
Visitwise mean changes from baseline by treatment group and time of last observation in the high-dropout large data set.

baseline score-by-visit interaction. An unstructured (co)variance structure was used to model the within-patient errors. The Kenward–Roger approximation was used to estimate denominator degrees of freedom. Significance tests were based on least-squares means using two-sided $\alpha = 0.05$ (two-sided 95% confidence intervals). Analyses were implemented using SAS PROC MIXED (SAS, 2013). The primary comparison was the contrast between treatments at Week 8.

Results from the primary analyses are summarized in Table 25.2. In the high-dropout data set, the advantage of drug over placebo in mean change from baseline to Week 8 was 2.29 (SE = 1.00, p = 0.024). The corresponding results in the low-dropout data set were 1.82 (SE = 0.70, p = 0.010). The standard error for the difference in LSMEANS at Week 8 in the high-dropout data set was nearly 50% larger than in the low-dropout data set. This was due to the lower variability and higher completion rate in the low-dropout data set.

Given that the mechanism by which the likelihood-based repeated measures analysis estimates what would have happened if patients had not discontinued early is via the residual correlations, it is important to assess sensitivity of results to the correlation structure. Unstructured covariance matrices were used for the primary analyses. Correlations and (co)variances from the primary analyses of the high- and low-dropout data sets are summarized in Table 25.3.

Treatment contrasts from more general and more parsimonious covariance structures, with and without use of the sandwich estimator, are summarized

TABLE 25.2

Visitwise LSMEANS and Contrasts for HAMD17 from the Primary Analyses of the High- and Low-Dropout Data Sets

| | | | LSMEANS | LSMEAN | |
	Placebo	Drug	Difference[a]	Standard Error	p-Value
High Dropout					
Week 1	−1.89	−1.80	−0.09	0.64	0.890
Week 2	−3.55	−4.07	0.52	0.81	0.518
Week 4	−4.87	−6.20	1.32	0.86	0.125
Week 6	−5.51	−7.73	2.22	0.93	0.019
Week 8	−5.94	−8.24	2.29	1.00	0.024
Low Dropout					
Week 1	−2.22	−1.76	−0.45	0.38	0.235
Week 2	−4.93	−4.89	−0.04	0.56	0.943
Week 6	−7.79	−8.31	0.52	0.61	0.392
Week 4	−9.45	−10.69	1.25	0.66	0.060
Week 8	−10.57	−12.39	1.82	0.70	0.010

[a] Advantage of drug over placebo. Negative values indicate an advantage for placebo.

TABLE 25.3

Covariance and Correlation Matrices from the Primary Analyses of the High- and Low-Dropout Data Sets

Week	High Dropout					Low Dropout				
	1	2	4	6	8	1	2	4	6	8
(Co)variances										
1	20.16					7.22				
2	14.05	29.86				4.79	15.50			
4	11.55	19.74	32.09			3.87	11.54	18.07		
6	10.70	18.56	25.27	35.10		3.74	9.96	14.65	20.59	
8	11.51	17.67	22.57	30.88	39.55	2.23	7.03	10.73	16.61	22.82
Correlations										
1	1.000					1.000				
2	0.573	1.000				0.453	1.000			
4	0.454	0.638	1.000			0.339	0.689	1.000		
6	0.402	0.573	0.753	1.000		0.307	0.558	0.760	1.000	
8	0.408	0.514	0.634	0.829	1.000	0.174	0.374	0.528	0.766	1.000

in Table 25.4. In the high-dropout data set, an unstructured matrix common to both treatment groups that was used as the primary analysis provided the best fit, thereby supporting results from the primary analysis as valid. However, the potential importance of choice of covariance structure with this high rate of dropout can be seen in the comparatively large range in treatment contrasts (1.69–2.29), standard errors, and p-values across the covariance structures.

In the low-dropout data set, the range in treatment contrasts across the covariance structures (1.76–1.85) was approximately sixfold smaller than in the high-dropout data set, and all structures yielded a significant treatment contrast. Separate unstructured matrices by treatment group yielded the best fit, with the second-best fit being from a single unstructured matrix that was specified as the primary analysis. The consistency of results across covariance structures indicates conclusions were not sensitive to correlation structure, thereby supporting validity of the primary analysis.

Sensitivity of the primary analysis to the assumption that missing data arose from a missing at random mechanism was assessed using a marginal delta adjustment analysis.

In the present application, $m = 100$ imputations was shown to stabilize results and therefore each delta adjustment analysis used 100 imputed data sets. Results from applying marginal delta adjustment to the high- and low-dropout data sets are summarized in Table 25.5. In the marginal approach, the delta adjustment at one visit does not influence imputed values at other visits. Results from

TABLE 25.4

Treatment Contrasts from Alternative Covariance Matrices from the Primary Analyses

Structure[a]	AIC	Endpoint Contrasts	Standard Error	p-Value
High Dropout				
UN	4679.82	2.29	1.00	0.024
UN EMPIRICAL	4679.82	2.29	0.97	0.020
TOEPH	4684.44	2.10	0.91	0.023
TOEPH EMPIRICAL	4684.44	2.10	0.92	0.023
TOEPH GROUP=TRT	4689.88	1.82	0.91	0.048
UN GROUP=TRT	4692.05	1.96	1.00	0.053
CSH	4735.81	1.86	0.93	0.047
CSH EMPIRICAL	4735.81	1.86	0.91	0.041
CSH GROUP=TRT	4739.34	1.69	0.93	0.070
Low Dropout				
UN GROUP=TRT	4861.70	1.85	0.703	0.009
UN	4867.68	1.82	0.699	0.010
UN EMPIRICAL	4867.68	1.82	0.666	0.007
TOEPH GROUP=TRT	4888.93	1.82	0.647	0.005
TOEPH	4897.89	1.79	0.649	0.006
TOEPH EMPIRICAL	4897.89	1.79	0.662	0.006
CSH	5030.40	1.76	0.705	0.013
CSH EMPIRICAL	5030.40	1.76	0.667	0.008
CSH GROUP=TRT	5031.92	1.80	0.708	0.011

Note: [a] UN = unstructured; toeph = heterogeneous toeplitz; CSH = heterogeneous compound symmetric; GROUP = TRT means that separate structures were fit for each treatment group; empirical means that empirical (sandwich-based) estimators of the standard error were used rather than the model-based standard errors.

using delta = 0, that is no adjustment, the MAR result, are included for reference. The MAR (delta = 0) results differ slightly from the primary results in Table 25.2 because the primary result was generated using direct likelihood and the delta = 0 results in Table 25.5 are based on multiple imputation.

Deltas were progressively increased in the tipping-point format to identify the magnitude of delta needed to overturn the primary results. Delta = 2 was a particularly informative choice because that delta was approximately equal to the treatment effect and therefore is a useful reference point in gauging plausibility of a departure from MAR (Permutt, 2015b).

Using delta = 2, the endpoint contrast was significant in the low-dropout data set, but significance was lost in the high-dropout data set. In fact, delta = 1 was sufficient to overturn significance in the high-dropout data set. In the low-dropout data set, delta = 7 was required to overturn significance.

TABLE 25.5

Results from Marginal Delta-Adjustment Multiple Imputation – Delta Applied on Last Visit to Active Arm Only

Value of Delta Adj	Low-Dropout Data Set			High-Dropout Data Set		
	Endpoint Contrast	Standard Error	*p*-Value	Endpoint Contrast	Standard Error	*p*-Value
0	1.86	0.70	0.008	2.27	1.12	0.042
1	1.79	0.70	0.011	1.96	1.13	0.083
2	1.71	0.70	0.015	1.64	1.14	0.151
3	1.64	0.71	0.201			
4	1.57	0.71	0.027			
5	1.50	0.72	0.037			
6	1.42	0.72	0.049			
7	1.35	0.73	0.065			

Increasing the delta applied to the endpoint visit had a consistent and therefore predictable effect on the endpoint contrasts. For each 1-point increase in delta, the endpoint contrast was decreased by approximately 0.08 points in the low-dropout data set and by approximately 0.30 points in the high-dropout data set. As discussed in Chapter 23, this systematic change is intuitive in that 8% of the values were missing for Treatment 2 at endpoint in the low-dropout data set and 30% of the values were missing for Treatment 2 in the high-dropout data set.

As discussed in Chapter 23, the effect of a marginal delta adjustment on the endpoint contrast can be analytically determined as follows:

$$\text{Change to the endpoint contrast} = \Delta \times \pi$$

where π = the percentage of missing values and Δ = the marginal delta adjustment.

These results help reinforce the straightforward and easy-to-understand nature of marginal delta adjustment. First, there is a direct correspondence between the fraction of missing values and the sensitivity of results to departures from MAR. If π is cut in half, Δ will have half the effect on the endpoint contrast. Moreover, progressively increasing delta results in predictable and easy-to-understand changes in the endpoint contrasts that make for a useful stress test to ascertain how severe departures from MAR must be in order to overturn inferences from the MAR result.

Results from reference-based imputation analyses to implement the hypothetical strategy of assuming patients who discontinued active drug stayed in the study on placebo are summarized in Table 25.6. The jump-to-reference (J2R) approach was used in which the benefit of the drug immediately disappears after discontinuation. Analyses were conducted using the SAS macros

TABLE 25.6

Results from Reference-Based Multiple Imputation of the High- and Low-Dropout Data Sets

	Placebo	LSMEAN Changes Drug	Endpoint Contrast	Standard Error	*p*-Value
High Dropout					
MAR	−5.95	−8.24	2.29	1.00	0.024
J2R	−5.97	−7.57	1.60	0.99	0.110
Low Dropout					
MAR	−10.56	−12.40	1.84	0.70	0.009
J2R	−10.55	−12.26	1.71	0.70	0.016

made available freely to the public by the Drug Information Association's (DIA's) Scientific Working Group for missing data at www.missingdata.org.uk.

As expected, treatment contrasts from J2R were smaller than from an otherwise similar MAR-based analysis. In the high-dropout data set, the endpoint contrast from J2R was 1.60 (SE = 0.99, $p = 0.110$). In the low-dropout data set, the endpoint contrast from J2R was 1.71 (SE = 0.70, $p = 0.016$). The difference from MAR was approximately sixfold smaller in the low-dropout data set than in the high-dropout data set. Statistical significance was preserved in the low-dropout data set but not for the high-dropout data set.

The preceding examples from the high- and low-dropout data sets illustrated some fundamental points in analyzing longitudinal data. Perhaps the most important aspect of these illustrations was the benefit from lower rates of missing data. Sensitivity analyses were used to aid understanding of the degree to which departures from MAR for the unobserved data could alter inferences from the primary analysis. Results were much more robust to missing data assumptions in the low-dropout data set.

In the high-dropout data set, the possibility of plausible departures from MAR overturning the primary result could not be entirely ruled out. In the low-dropout data set, inferences were robust to even the largest plausible departures from MAR.

26

Analytic Case Study Based on the ACTG 175 HIV Trial

26.1 Introduction

The ACTG 175 study was a randomized, double-blind trial comparing nucleoside monotherapy versus combination therapy in human immuno-deficiency virus (HIV)-infected adults with CD4 T-lymphocyte (CD4) cell counts from 200 to 500 per cubic millimeter (Hammer et al., 1996). When the study was conducted in the early 1990s, an antiretroviral therapy with zidovudine (ZID) had shown beneficial effects on survival, the incidence of opportunistic infections, and the rate of disease progression in patients with advanced type 1 HIV. However, its effectiveness decreased with pro-longed use, and regimens combining ZID with other antiretroviral agents yielded better and more durable outcomes in advanced HIV patients.

The ACTG 175 study investigated the effects of monotherapies and combination therapies in adult patients with less advanced HIV disease; that is, CD4 cell counts between 200 and 500 per cubic millimeter within 30 days before randomization, no history of acquired immunode-ficiency syndrome (AIDS) defining illness other than minimal mucocu-taneous Kaposi's sarcoma, a Karnofsky performance score of at least 70, and acceptable laboratory results. The key clinical outcomes of interest in HIV studies at the time were ≥50% decline in the CD4 cell count, events indicating progression to AIDS, and mortality. These outcomes had often been evaluated as composite time-to-event endpoints assessing the time to the first of these events.

The ACTG 175 study randomized 2467 HIV-1-infected patients, with or without prior antiretroviral therapy, to one of the four daily regimens involving three previously approved antiretroviral agents: 600 mg of ZID, 600 mg of ZID plus 400 mg of didanosine (ZID+DID), 600 mg of ZID plus 2.25 mg of zalcitabine (ZID+ZAL), and 400 mg of DID. Randomization was stratified by the length of prior antiretroviral therapy. Patients who experienced a ≥50% decline in the CD4 cell count or an AIDS event were offered a combination therapy as a form of rescue, without revealing their initial treatment assignment. More specifically, patients originally randomized to one of the monotherapies, ZID or DID, were randomized in a blinded fashion to one of the combination therapies, ZID+DID or ZID+ZAL, while patients who had been randomized to combination therapy remained on the same therapy.

Patients were followed until the global study termination date (when a planned total number of events for the primary endpoint was accrued) and the last patient enrolled had an opportunity to complete 24 months of treatment. Patients were followed regardless of whether they remained on the randomized treatment or discontinued it prematurely and regardless of the initiation of rescue. Patients had their CD4 counts measured at baseline, week 8, and every 12 weeks thereafter.

The primary endpoint was time to the first occurrence of ≥50% decline in the CD4 cell count, an AIDS event, or death. A treatment policy strategy was used for all intercurrent events (ICEs). One of the secondary endpoints was time to the first occurrence of an AIDS event or death, also under a treatment policy strategy. The study manuscript also provided descriptive summaries of mean changes from baseline in CD4 counts over time. Achieving higher CD4 counts over time is associated with better outcomes (Hoffman et al., 2010).

This study is used to illustrate some concepts and methods discussed in this book. The goal is not to derive new interpretations regarding the efficacy of the study treatments or to provide a clinical justification for alternative estimands. Rather the purpose is to illustrate how alternative approaches could be implemented. In the next section, several estimands are discussed. They are not the ones that address the actual primary objective in the ACTG study but rather illustrations for various types of objectives that may be pursued in a study with a similar design and types of ICEs. To streamline the discussion and analyses, focus is on a subset of patients that compares one of the study monotherapies, the ZID treatment group, to a combined group composed of the two combination therapies: ZID+DID and ZID+ZAL.

26.2 Study Details

The rate of premature treatment discontinuation was 62% in the ZID monotherapy group and 55% in the combination therapy group. In the monotherapy and combination therapy groups, 11% and 15% discontinued their initial randomized treatment prematurely due to toxicity, respectively, and 3% and 2% died or discontinued treatment due to lack of efficacy in the two groups, respectively. Other treatment discontinuations were patient decisions for reasons including low-grade toxic reactions, declining CD4 cell counts, the desire to seek other therapies, and study burden.

Moreover, 30% received rescue in the monotherapy treatment group and 17% were re-randomized to rescue after meeting the rescue criteria in the combination treatment group although they remained on their original treatment per the study design. The rescue rates were 21% and 10% prior to Month 24 in the two treatment groups, respectively. Some patients (8% and 7% in the two treatment groups, respectively), met the criteria for rescue, but did not initiate it and instead discontinued study treatment. Figure 26.1 presents a Kaplan–Meier plot of time to meeting the protocol-defined rescue criteria in the two treatment groups, where a significant difference between the treatment groups is evident.

Among patients included in our case study, 22% were lost to follow-up in both the monotherapy and combination therapy group. Therefore, more than half of the patients who discontinued treatment prematurely remained on study and continued to be followed. The rates of death were similar during the first 2 years post-randomization but started to diverge thereafter in the two treatment groups (see Figure 26.2).

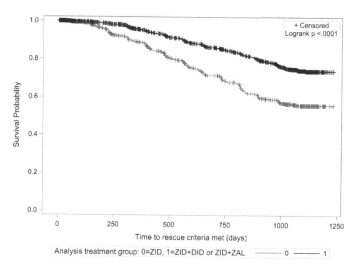

FIGURE 26.1
Kaplan–Meier plot for time to meeting protocol-defined criteria for rescue.

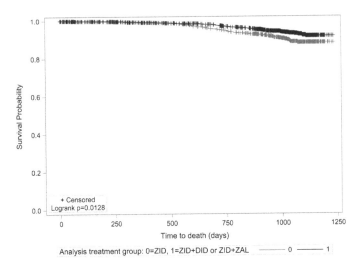

FIGURE 26.2
Kaplan–Meier plot for time to death.

26.3 Estimands and Estimators for the Case Study

26.3.1 Introduction

In this section, estimands and their corresponding estimators for two types of endpoints are discussed: continuous, based on CD4 counts over 24 months, and time to event (an AIDS event or death, whichever comes first). With each type of endpoint, several objectives and the estimands that support these objectives are considered. In each case, the study development process chart outlined in Chapter 3 is followed, except that retention strategies and sample size are not discussed because they are not relevant to this reanalysis.

Each estimand considered in this case study is intended for the same decision maker: a prescribing physician who seeks to select a treatment for his or her patients. Therefore, the decision maker is not identified in the definition of each estimand and the item *"a: Identify decision maker"* is skipped. Considerations are somewhat different for each estimand and are elaborated in the item *"1b. Define objective."*

26.3.2 Estimands for the Continuous Endpoint of Change from Baseline in CD4

Two estimands are considered for the objective of comparing treatment regimens in terms of their effect on the CD4 counts at 24 months after initiation of the study treatment, with the key difference being the treatment regimens and population under evaluation. For both estimands, the study treatment includes following the current clinical practice regarding treatment discontinuation due to toxicity or switching to other nonstudy therapies per the physician and patient choice.

Estimand 26.1 "Treatment Policy and Composite Strategy Estimand for Median Change from Baseline to Month 24 in CD4 Count" in All Randomized Patients

Definition

    1b. Define objective: The main objective is to inform clinical decision making on selecting a treatment strategy in adult patients with a less advanced HIV disease based on change from baseline in CD4 counts at 24 months after treatment initiation. The intent is to evaluate benefits of initiating treatment with a combination therapy versus a monotherapy with an option of switching to a combination therapy if predefined criteria for unsatisfactory clinical response are met.

    2a. Identify possible ICEs: Treatment regimens that may occur in this trial in either a planned or unplanned manner are summarized in Table 26.1. Scenario 1 represents adherence to the randomized treatment through the double-blind treatment period, without any ICEs. (Adherence does not mean 100% compliance, but rather continuing with the initially randomized treatment for 24 months.) Treatment changes, such as switch to protocol-designed rescue therapy, to no treatment, or to an alternative therapy are scenarios 2, 3, and 4, respectively. These are ICEs that may occur at any time. Use of other antiretroviral agents concomitantly with the study treatments is prohibited. Patients may also die during the study while following any of the above-mentioned treatment regimens.

    2b. Define treatment regimen under evaluation: The intended treatment regimen is the initially randomized treatment (ZID monotherapy or combination therapy with either DID or

TABLE 26.1

Treatment Regimens and Intercurrent Events Anticipated in the ACTG-175 Trial over 24 Months Post-randomization

Scenario	Treatment Regimens	Events
1	Z	Death
2	Z→R	
3	Z→P	
4	Z→R→P	

Note: Z = randomized treatment (monotherapy or combination therapy); R = rescue therapy (monotone therapy switch to a combination therapy); P = premature discontinuation of study treatment with or without a switch to an alternative therapy.

zalcitabine, ZID + [DID or ZAL]) and subsequent changes in treatment such as early treatment discontinuation, switch to another nonstudy treatment, or rescue from monotherapy to combination therapy per the protocol-defined rescue procedure, over 24 months post-randomization.

2c. *Define estimand*: The estimand is defined as follows, specifying the four elements (population, outcome, handling of relevant ICEs, population-level summary of treatment effect) as outlined in ICH E9(R1):

   a. The treatment effect is estimated for all randomized patients in the population of HIV-infected adult patients as defined by the protocol inclusion/exclusion criteria.

   b. Efficacy is measured using the endpoint of change from baseline to Month 24 of the double-blind study period in CD4 counts.

   c. All ICEs leading to changes in treatment, such as premature discontinuation of the randomized treatment, with or without a switch to alternative therapies, and initiation of the protocol-defined rescue are handled by a treatment policy strategy. The ICE of death is handled with a composite strategy where death is a treatment failure.

   d. The changes from baseline to Month 24 in CD4 counts is compared using the difference between groups in median change from baseline.

3a. *Data needed for estimand*: The CD4 count data necessary for this estimand are those observed at Month 24 post-randomization regardless of treatment changes. Therefore, the data after any ICE, except death, must be collected through Month 24. Survival status and date of death are also needed.

4a. *Main estimator*: An estimator aligned with the estimand needs to handle continuous values of change from baseline to Month 24 in CD4 counts of patients who survive to Month 24 together with the treatment failure outcome of patients who die prior to Month 24. A nonparametric rank-based approach wherein death is assigned the worst rank; that is, worse than any rank of a surviving patient based on the CD4 endpoint value. All deaths, regardless of timing, are assigned the same worst rank. The hypothesis test is conducted using a rank-based analysis of covariance (ANCOVA) adjusting for baseline CD4 counts and length of prior antiretroviral therapy. The treatment effect is estimated as the difference between group medians using quantile regression adjusting for the same baseline covariates.

4b. *Missing data assumption*: Missing data is expected due to patient withdrawal prior to Month 24. Missing outcomes for these patients (survival status at Month 24 and CD4 count of surviving patients) are imputed assuming that the outcomes would follow a distribution similar to that for patients in their treatment group who discontinued the study treatment prematurely but remained in the study through Month 24, conditioning on baseline characteristics and post-baseline CD4 counts. This assumption is aligned with the estimand and ICE strategies in element C.

4c. *Sensitivity estimators*: Sensitivity analysis assess the robustness of conclusions from the main estimator to the missing data assumption described in 4b by assuming that patients with missing CD4 values at Month 24 have worse outcomes compared to patients remaining in the study. A sensitivity analysis based on this assumption is conducted by assigning ranks to surviving patients with missing Month 24 CD4 counts that are worse than any rank of patients with available CD4 counts but better than the rank designated for death prior to Month 24. Analysis methods will be the same as described in 4a (rank-based ANCOVA and quantile regression) but applied to the alternative ranking described here. Moreover, as a supportive summary of treatment effect, the "win ratio" is calculated, both under the ranking as in the main estimator and in the sensitivity analysis.

Results are presented in Table 26.2 for the "Treatment policy and composite strategy estimand for change from baseline to Month 24 in CD4 count." The hypothesis of no difference between the monotherapy and combination therapy is rejected at the 0.05 two-sided significance level for both the main analysis and the sensitivity analysis. In the main analysis, the difference between endpoint medians of the combination therapy versus the monotherapy is 77.2, 95% CI = (63.74, 93.02), indicating that patients randomized to the combination therapy had a greater increase in CD4 counts from baseline to Month 24 compared to patients randomized to monotherapy. The sensitivity analysis shows that these conclusions hold under the more pessimistic assumptions about the individual CD4 counts of surviving patients with missing values of the CD4 count at Month 24. In fact, the difference between treatments is larger under the sensitivity assumption. Under this assumption, the missing

TABLE 26.2

Results of Analyses for the Estimand "Treatment Policy and Composite Strategy Estimand for Change from Baseline to Month 24 in CD4 Count"

Statistic	Main Estimator	Sensitivity Estimator
p-Value, rank ANCOVA	<0.0001	<0.0001
Difference in medians (95% CI), quantile regression	77.2 (63.74, 93.02)	102.8 (66.95, 138.61)
Win ratio (95% CI)	1.8 (1.60, 2.07)	1.5 (1.36, 1.68)

Note: Number of patient included in the analysis: Monotherapy $N = 619$, Combination $N = 1228$.

cases are shifted to the worse end of the distribution, leaving median estimates to be more impacted by the observed values, which exhibit a larger treatment effect.

The conclusion about more favorable CD4 outcomes after starting a combination therapy is also supported by the result from the win ratio analysis that favors the combination therapy group with the win ratio estimate of 1.8, 95% CI = (1.60, 2.07) under the missingness assumption of the main analysis. The win ratio is interpreted as the odds of having a larger improvement in CD4 counts for combination therapy compared to monotherapy.

The win ratio results from the sensitivity analysis are also consistent with the overall conclusion. However, contrary to the pattern noted with the quantile regression estimate of the difference in medians, the win ratio from the sensitivity analysis is smaller than from the main analysis. The win ratio estimate is influenced by the entire distribution rather than reflecting the difference in specific quantiles.

Estimand 26.2 "Treatment Policy and Principal Stratification Strategy Estimand for Median Change from Baseline to Month 24 in CD4 Count"

Definition

  1b. Define objective: The main objective is to inform clinical decision making for treatment strategy selection in adult patients with less advanced HIV disease. Contrary to the objective of Estimand 26.1, the focus here is on evaluating benefits of following a combination therapy for 24 months versus a monotherapy for 24 months in patients who would survive without requiring rescue, per the protocol-defined rescue criteria, through Month 24. In other words, the objective is to evaluate whether the combination therapy provides any advantage in terms of improved CD4 counts in patients who would respond sufficiently well to both combination therapy and monotherapy, where sufficient response is defined by the same criteria as rescue requirements and survival. This is of interest in order to reduce unnecessary treatments.

  2a. Identify possible ICEs: The ICEs that are expected after the initial randomized treatment assignment are the same as described for Estimand 26.1 and summarized in Table 26.1.

  2b. Define treatment regimen under evaluation: The treatment regimen intended for evaluation is the initially randomized treatment (ZID monotherapy or combination therapy with either DID or zalcitabine, ZID + [DID or ZAL]) over 24 months without rescue.

  2c. Define estimand: The estimand is defined as follows, specifying the four elements as outlined in ICH E9(R1):

    a. The treatment effect is estimated for the population of HIV-infected adult patients as defined by the protocol inclusion/exclusion criteria who would survive without requiring rescue, per the protocol-defined rescue criteria, through Month 24 on any treatment.

    b. Efficacy is measured using change from baseline to Month 24 in CD4 counts.

    c. The ICEs of death and initiation of the protocol-defined rescue prior to Month 24 will be handled by the principal stratification strategy. Other ICEs leading to changes in treatment, such as premature discontinuation of the randomized treatment without initiation of rescue or death prior to Month 24 will be handled using the treatment policy strategy.

    d. The median of changes from baseline to Month 24 in CD4 counts is estimated for each treatment group, and the combination therapy group is compared to the monotherapy group using difference in group medians.

3a. Data useful for estimand: The data necessary for this estimand are date of death and date when the rescue criteria are met or censoring dates for such events for all randomized patients; baseline patient and disease characteristics predictive of the outcomes listed in (i) for all randomized patients; and CD4 counts at Month 24 for patients who survived to Month 24 without rescue.

4a. Main estimator: An estimator aligned with the estimand models the membership of patients in the principal stratum of interest. The principal strata in this context are defined based on the ICEs of death or meeting rescue criteria prior to Month 24 post-randomization (see Chapter 24, Section 24.4, for a general discussion of this type of stratification). Let $S(Z) \in \{0,1\}$ denote whether such an ICE occurs or not if a patient follows therapy Z, where $Z = 0$ represents the monotherapy and $Z = 1$ represents the combination therapy.

The four principal strata that can be defined based on these ICEs are defined in Table 26.3: the columns of this table represent the patient status with respect to the ICE if the patient follows treatment $Z = 1$, and the

TABLE 26.3

Principal Strata Defined Based on the ICE of Death or Meeting Rescue Criteria Prior to Month 24

	$S(1) = 1$	$S(1) = 0$
$S(0) = 1$	Rescues/dies prior to Month 24 on any treatment	Insufficient response to or death on monotherapy only prior to Month 24
$S(0) = 0$	Insufficient response to or death on combination therapy only prior to Month 24	Survives without rescue through Month 24 on any treatment

rows represent the patient status if the patient follows treatment $Z = 0$. As defined in part A of the estimand definition (item 2c), the focus of this estimand is on the principal stratum of patients who "survive w/o rescue through Month 24 on any treatment." The abbreviation "SWoR–any" is used for this principal stratum.

The membership in "SWoR–any" cannot be observed based on data collected in this study because each patient is randomized to one treatment – either monotherapy or combination therapy. Therefore, the membership of each patient in "SWoR–any" is modeled in the framework of analysis with missing data (see Chapter 24, Section 24.6). More specifically, with respect to the principal stratum membership:

> For patients randomized to $Z = 0$ (monotherapy): S(0) is observed and S(1) is missing.
> For patients randomized to $Z = 1$ (combination therapy): S(1) is observed and S(0) is missing.

To determine each patient's membership in "SWoR–any," first predict and impute each patient's missing status, (0) or (1). The imputation of the missing status and analysis of the endpoint within the principal stratum "SWoR–any" is accomplished using multiple imputation as follows:

> For each treatment $Z \in \{0,1\}$, fit a Bayesian imputation model to observed data (S^{obs}, Z, X) in order to predict S from (Z, X), where X is a vector of baseline covariates. Take M draws from a posterior distribution of model parameters.
> For each $m = 1, \ldots, M$:
> For each randomized patient with treatment indicator Z_i, impute missing status of ICE occurrence on the other treatment, that is, the treatment to which the patient was not randomized, $\hat{S}^m(1 - Z_i)$.
> Form a principal stratum I_m of patients who would not experience S regardless of the treatment:

> For a patient i randomized to $Z = 0$: $i \in I_m$ if $(S_i^{obs}(0) = 0 \ \& \ \hat{S}_i^m(1) = 0)$
>
> For a patient i randomized to $Z = 1$: $i \in I_m$ if $(S_i^{obs}(1) = 0 \ \& \ \hat{S}_i^m(0) = 0)$

> Estimate the treatment effect $\hat{\delta}_m(I_m)$ as the difference between medians of change from baseline to Month 24 in CD4 counts in all patients $i \in I_m$.
> Compute an overall estimate of treatment effect $\hat{\delta}(I)$ using Rubin's rule by combining estimates $\hat{\delta}_m(I_m)$ from all imputed data sets $m = 1, \ldots, M$.

The multiple imputation model is a piecewise exponential survival model for time-to-event data modeling time to death or rescue, whichever comes first, based on a set of baseline covariates. The imputed status $\hat{S}^m(z)$ is considered 0 if the imputed time of event (death or rescue) is after Month 24 or is censored.

Within the principal stratum "SWoR–any" formed from each imputed data set, that is, a subset of patients I_m, change from baseline to Month 24 in CD4 counts is analyzed as follows. The hypothesis test is carried out using rank-based ANCOVA adjusting for baseline CD4 counts and the length of prior antiretroviral therapy. The treatment effect is estimated as the difference between group medians using quantile regression adjusting for the same baseline covariates.

4b. *Missing data assumption*: In the context of this estimand, missing data is expected due to some surviving patients withdrawing from the study prior to Month 24 without meeting criteria for rescue. Missing ICE status of these patients (i.e., survival and rescue status at Month 24) on the randomized treatment to which they were assigned is imputed from the same multiple imputation model for time to the ICE as described in the main estimator. In other words, the ICE status on both treatments is imputed for such patients.

For patients who are predicted to belong to the "SWoR–any" stratum but have a missing value of the CD4 count at Month 24, CD4 count is imputed assuming that it follows a distribution similar to that for patients with available Month 24 CD4 counts in their treatment group and in the "SWoR–any" stratum, conditioning on baseline CD4 counts, the length of prior antiretroviral therapy, and available post-baseline CD4 counts.

4c. *Sensitivity estimators*: Sensitivity analysis is conducted to assess the robustness of conclusions from the main estimator to the missing data assumption described in 4b. It can be stress tested by assuming that patients withdrawing from the study would have worse CD4 counts compared to patients remaining in the study. A sensitivity analysis based on this assumption is conducted by assigning ranks to patients with missing Month 24 CD4 counts that are worse than any rank of patients with available CD4 counts. Analysis methods are the same as described in 4a but applied to the alternative ranking described here. Moreover, as a supportive summary of treatment effect, the win ratio will be calculated, both under the ranking as used in the main estimator and sensitivity analysis.

Membership in the "SWoR–any" stratum was modeled based on the following baseline factors: age, race, weight, baseline CD4 count, presence of hemophilia, homosexual activity, history of intravenous drug use, Karnofsky's performance score, symptomatic/asymptomatic status, antiretroviral therapy naive/experienced status, number of days of prior antiretroviral therapy, ZID naive/experienced status, and number of days of prior ZID therapy.

Results for the estimand "Treatment policy and principal stratification strategy estimand for change from baseline to Month 24 in CD4 count" are presented in Table 26.4. The average number of patients included in the principal stratum was 406 and 808 for the monotherapy and combination therapy groups, respectively, which is 66% of all randomized patients in each treatment group. Equal percentages in the two treatment groups are expected because the principal stratum membership is independent of the randomized treatment assignment.

TABLE 26.4

Results of Analyses for the Estimand "Treatment Policy and Principal Stratification Strategy Estimand for Change from Baseline to Month 24 in CD4 Count"

Statistic	Main Estimator	Sensitivity Estimator
p-Value, rank ANCOVA	<0.0001	<0.0001
Difference in medians (95% CI), quantile regression	57.3 (36.99, 77.56)	76.1 (36.17, 116.08)
Win ratio	1.7 (1.43, 2.03)	1.5 (1.26, 1.69)

Note: Average number of patients included in the principal stratum (% of all randomized): monotherapy $N = 406(66\%)$, combination $N = 808(66\%)$.

The hypothesis of no difference between the monotherapy and the combination therapy in median change from baseline to Month 24 in CD4 counts in the principal stratum "SWoR–any" is rejected at the 0.05 two-sided significance level using rank-based ANCOVA, both for the main analysis and sensitivity analysis. In the main analysis, the difference between endpoint medians of the combination therapy versus the monotherapy is 57.3, 95% CI = (36.99, 77.56), indicating that patients on the combination therapy had a larger median increase in CD4 counts from baseline to Month 24 compared to patients on the monotherapy. The treatment difference was smaller in this patient subset compared to the overall study population as estimated by Estimand 26.1, which could be expected because there may be less room for improvement in a subset of relatively healthier patients. The sensitivity analysis shows that these conclusions hold under the alternative missingness assumptions. As in results for estimand 26.1, the difference between medians is larger under the sensitivity assumptions compared to the main estimator.

The overall conclusion is also supported by the result from the win ratio analysis that favors the combination therapy with the win ratio estimate of 1.7, 95% CI = (1.43, 2.03). The win ratio result from the sensitivity analysis is consistent with the main analysis overall, but the estimated win ratio is somewhat smaller.

26.3.3 Estimands Based on the Time-to-Event Outcomes

This section includes three estimands with the objective of evaluating the treatment effect on time to the first AIDS event or death. The first estimand is based on a treatment policy strategy and the two others – on a principal stratification strategy.

> **Estimand 26.3** "Treatment Policy Estimand for Time to AIDS or Death in All Randomized Patients"
>
> > *1b. Define objective:* The objective is to evaluate the risk of AIDS or death for patients starting treatment with a monotherapy

with an option of switching to a combination therapy later, as described in Section 26.1, compared with patients who started and continued on the combination therapy.

2a. *Identify possible ICEs*: The same as for Estimand 26.1, except the death is part of the outcome.

2b. *Define treatment regimen under evaluation:* The treatment regimen intended for evaluation is the initially randomized treatment (ZID monotherapy or combination therapy ZID + [DID or ZAL]) and subsequent changes in treatment such as early treatment discontinuation, switch to another nonstudy treatment, or rescue from monotherapy to combination therapy as per the protocol-defined rescue procedure.

2c. *Define estimand*:

   a. The population is the same as for Estimand 26.1.

   b. Efficacy is evaluated based on time to the first AIDS event or death.

   c. Specific ICEs that may result in changes in treatment (50% reduction in CD4 counts and premature discontinuation) are handled by the treatment policy strategy. Both AIDS (which also may result in treatment change from monotherapy to the combination therapy as rescue) and death are included in the outcome.

   d. The effect of the treatment regimen starting with a combination therapy versus starting with the monotherapy will be summarized with the hazard ratio for the outcome event defined in part B.

3a. *Data needed for estimand*: The data necessary for this estimand include time to event or censoring and event or censoring status, along with covariates used in the analysis model that are collected regardless of occurrence of any ICEs.

4a. *Main estimator*: The estimator aligned with the estimand is a covariate-adjusted estimate of the treatment coefficient in the proportional hazards Cox regression model for time to event (COX PH).

4b. *Missing data assumption*: For the primary analysis, patients who withdraw from the study without experiencing the event (AIDS or death) are handled by the COX PH estimation procedure based on the assumption of censoring at random (CAR).

4c. *Sensitivity estimators*: The CAR assumption is stress tested using a multiple imputation-based procedure described in Lipkovich, Ratitch, and O'Kelly (2016). Event times for patients who are censored are imputed using a Bayesian piecewise exponential proportional hazards model (PEPHM) assuming a larger event hazard for patients who discontinued from the study within 24 months after randomization. The imputed data will be analyzed with the same Cox regression model, as used for the main analysis. More specifically, the sensitivity analysis is implemented with the following steps.

- *Fit Bayesian PEPHM* with the number of intervals $k = 3$ to the analysis data set for time to event separately by treatment group. The interval cutoffs are determined by the procedure to ensure an approximately equal number of events per interval within each treatment group.
- *Sample from the posterior distribution* of model parameters (interval-specific hazards and regression coefficients, $m = 50$ draws).
- *Generate event times* for patients who discontinued from the study using imputation from the posterior predictive distribution of time to event for that patient's treatment group, with covariate-specific hazard functions multiplied by a treatment-specific sensitivity parameter: δ_0 or δ_1 for the initial mono- and combination therapies, respectively. For each draw from the posterior distribution, generate missing event times $\widetilde{T}_i$ for each censored patient starting from the time he or she discontinued from the study (C_i) using piecewise exponential hazards model (for details, see Lipkovich, Ratitch, and O'Kelly, 2016) and retain this value as imputed time to event or censor at the end of planned follow-up (T_{max}), if $\widetilde{T}_i > T_{max}$. (Here, T_{max} was set close to the planned follow-up time = 730 days as the planned follow-up was 2 years.) This results in generating $m = 50$ imputed data sets.
- *Analyze* each imputed data set using the PH COX analysis model and compute log (HR) and associated standard errors (SEs) for the treatment effect.
- *Combine* $m = 50$ estimated log(HR)s and associated SEs to produce an overall point estimate, confidence interval, and p-value using Rubin's combination rules.

A tipping point analysis is conducted by executing steps 1–5 for different values of the sensitivity parameters δ_0 and δ_1, for the mono- and combination therapy groups, respectively. For each setting, a p-value is computed, and values of delta identified when the null hypothesis is not rejected. For this case study, as is often the case, only δ_1 for the combination therapy (considered experimental in this context) is varied, implying increased hazard for patients following their withdrawal) while keeping $\delta_0 = 1$, corresponding to CAR for the monotherapy (considered control in this context).

The main analysis for time to AIDS or death, under the treatment policy strategy, is summarized in Table 26.5. Results show the superiority of combination therapy with a hazard ratio of 0.67 (0.52, 0.87). The same result is reproduced in the sensitivity analysis using multiple imputation of times to event for patients censored due to study withdrawal with sensitivity parameters $\delta = 1_0$, $\delta_1 = 1$ applied to the hazard function for the monotherapy and combination therapy, respectively. This is to be expected because both the main analysis of observed data and the analysis of multiply imputed data with the above-mentioned sensitivity parameter settings assume CAR.

When performing a sensitivity analysis by increasing δ_1 to assume an increased hazard for patients on combination therapy following study

TABLE 26.5

Results of Primary Analysis and Sensitivity Analysis for Estimand 26.3

		Sensitivity Estimator (MI + COX PH)	
Statistic	Main Estimator (COXPH)	$\delta_0 = 1, \delta_1 = 1$ (CAR)	$\delta_0 = 1, \delta_1 = 3$
p-Value,	0.0029	0.0025	0.105
HR, Confidence interval (95% CI)	0.67 (0.52, 0.87)	0.67 (0.51, 0.87)	0.80 (0.61, 1.05)

withdrawal, while keeping $\delta_0 = 1$, the point estimate for the covariate-adjusted hazard ratio becomes closer to unity reaching 0.8 for $\delta_1 = 3$ with a p-value of approximately 0.1. The p-value of 0.05 is reached for $\delta_1 = 2.4$. Therefore, to overturn the significance of the main analysis at a two-sided 0.05 level, more than twofold increase in the hazard in patients who withdrew from the study is needed.

Estimand 26.4 "Principal Stratification Estimand for Time to AIDS or Death (Not Needing Rescue on Any Study Treatment)"

1b. *Define objective*: The objective is to evaluate the risk of occurrence of AIDS or death for patients receiving a monotherapy as compared with that in patients receiving the combination therapy in a subpopulation of patients who would not experience a ≥50% worsening (i.e., decrease) from baseline in CD4 counts on any of these two treatments. Note that the target subpopulation is not the same as "those not needing rescue" based on the rescue conditions defined in the protocol, that is, a ≥50% decline from baseline in CD4 counts or an AIDS event. In this estimand, the occurrence of an AIDS event is part of the outcome (endpoint). Indeed, for some patients, an AIDS event may occur without a ≥50% worsening in CD4 counts. This estimand targets a subpopulation of patients who progress to AIDS or die without a prior symptomatic worsening in terms of severe CD4 count decline.

2a. *Identify possible ICEs*: The same as described in Table 26.1 (note death is part of the outcome).

2b. *Define treatment regimen under evaluation*: The treatment regimen intended for evaluation is the initially randomized treatment (ZID monotherapy or combination therapy with either DID or zalcitabine, ZID + [DID or ZAL]) without rescue.

2c. *Define estimand*:

a. The population is the principal stratum of patients who would not have a ≥50% reduction in CD4 count throughout the treatment regardless of the treatment group.

   b. Efficacy is evaluated based on time to the first AIDS event or death whichever comes first.
   c. Premature study treatment discontinuation is handled by the treatment policy strategy. Events resulting in rescue, such as $\geq 50\%$ worsening in CD4 counts will be handled by the principal stratification strategy, while AIDS events are included in the endpoint.
   d. The effect of the combination therapy versus the mono-therapy will be summarized with the hazard ratio for the outcome event defined in part B.

3a. Data needed for estimand: Data for this estimand include time to AIDS event, death, or *censoring*, censoring status, the event indicator for $\geq 50\%$ reduction in CD4 counts, along with covariates (the same as for Estimand 26.2). Post-randomization data after a premature discontinuation of study treatment is needed for this estimand.

4a. *Main estimator*: The same as in Estimand 26.3, except it is defined in the subpopulation $S(0) = 0, S(1) = 0$, as described in Table 26.6.

   A simple method for imputing missing ICE status ($S(0)$ for patients on the combination therapy and $S(1)$ for patients on monotherapy) is based on a Bayesian logistic regression model fitted to the observed data. The observed status $S(Z_i)$ of an ith patient assigned to Z_i who remained free of a $\geq 50\%$ CD4 count reduction until completion or withdrawal from the study or death is considered known: $S(Z_i) = 0$. Only counterfactual values $S(1 - Z_i)$ are imputed. The imputation and analysis proceed as described for Estimand 26.2, except a logistic imputation model is used with baseline covariates (with $m = 100$ imputations).

4b. *Missing data assumption*: Patients who are censored for the endpoint event (AIDS or death) are handled by the COX PH estimation procedure based on the assumption of CAR. Missing covariate values for a small number of patients with at least one missing covariate ($n = 15$) will be imputed using a multiple imputation procedure based on the MCMC method for normal distributions with observed status S included as covariate.

4c. *Sensitivity estimators*: The CAR assumption can be stress tested in the same way as in Estimand 26.3. For brevity, we omit these results.

The comparison of time to event (AIDS or death) for patients who started on combination therapy versus those who initiated the monotherapy is summarized in Table 26.7. The analysis is done on the principal stratum of patients predicted NOT to have a $\geq 50\%$ reduction in CD4 counts at any time on either of the two therapies. The proportion of patients allocated to a principal stratum varied from imputation to imputation resulting in the same average proportion of 65.1% in the monotherapy and combination therapy arms (403 and 799 patients, respectively). This is to be expected because stratum membership is by construction independent of treatment assignment, which is why

TABLE 26.6

Principal Strata Defined Based on the ≥50% Worsening of CD4 Counts at Any Time

	$S(1) = 1$	$S(1) = 0$
$S(0) = 1$	Worsening by ≥50% in CD4 counts at any time on any treatment	Worsening by ≥50% in CD4 counts at any time on monotherapy only
$S(0) = 0$	Worsening by ≥50% in CD4 counts at any time on combination therapy only	No ≥50% worsening in CD4 counts on both combination and monotherapy

TABLE 26.7

Summary of Results for the Principal Stratification Estimand 26.4. The Principal Stratum Is Composed of Subjects Predicted Not to Have a ≥50% Reduction (worsening) in CD4 Counts at Any Time on Either of the Two Therapies (approximately 65% across 100 imputed sets)

Statistic	Main Estimator
p-Value	0.3123
HR, Confidence interval (95% CI)	0.77 (0.47, 1.27)

valid causal inference can be conducted by summarizing treatment differences within each principal stratum, as if those were usual pretreatment covariates.

The results show that the advantage of combination therapy with respect to this endpoint is lost in this subpopulation of patients who would not have a symptomatic worsening regardless of treatment.

Estimand 26.5 "Principal Stratification Estimand for Time to AIDS or Death (Needing Rescue on Monotherapy Only)"

1a. *Identify decision maker*: The same as in Estimand 26.1.

1b. *Define objective*: The objective is to evaluate the risk of an AIDS event or death for patients starting a monotherapy and later switching to a combination therapy after a ≥50% decline in their CD4 counts as compared with patients who started and continued on the combination therapy, in a subpopulation of patients who would experience a ≥50% worsening in CD4 count on the monotherapy. (Note that this subpopulation may or may not experience a ≥50% worsening on a combination therapy.)

2a. *Identify possible ICEs*: The same as described in Table 26.1 (note death is part of the endpoint).

2c. *Define estimand*:

    a. The population is the principal stratum of patients who would have a ≥50% reduction in CD4 count at any time throughout the study after initiating treatment with the monotherapy.

TABLE 26.8

Summary of Results for the Principal Stratification Estimand 26.5.
The Principal Stratum Is Formed of Subjects Predicted to Have ≥50%
Reduction (worsening) in CD4 Counts at Any Time if Treated with
Monotherapy (approximately 19% of subjects averaged across
100 imputed sets)

Statistic	Main Estimator
p-Value	<0.0001
HR, Confidence interval (95% CI)	0.38 (0.25, 0.60)

    b. Efficacy is evaluated based on time to the first AIDS event
       or death whichever comes first.
    c. Same as for Estimand 26.4, with a distinction that the prin-
       cipal stratum of interest is different in this estimand.
    d. Same as for Estimand 3.

3a. Data needed for estimand: Same as for Estimand 26.4.
4a. Main estimator: The same as in Estimand 26.4, except it is defined
    in the subpopulation $S(0) = 1$, as described in Table 26.6.
4b. Missing data assumption: Same as for Estimand 26.4.
4c. Sensitivity estimators: Same as for Estimand 26.4.

The comparison of time to event (AIDS or death) for patients who started
on the combination therapy versus those initiating the monotherapy
and switching to combination therapy is summarized in Table 26.8.
The analysis is done on a principal stratum of patients who would have
≥50% reduction in CD4 counts if treated with monotherapy. The average
number of patients in this stratum (across 100 multiply imputed data sets)
was 159 in the monotherapy and 323 in the combination therapy group,
roughly corresponding to 19% of the total sample size. Interestingly, the
superiority of combination over initial monotherapy in this principal
stratum was substantially stronger compared to that in the overall
population: 0.38 (0.25, 0.6) versus 0.67 (0.52, 0.87), based on the primary
analysis.

References

Agresti A. (2002). *Categorical Data Analysis*. 2nd ed. New York: John Wiley & Sons.

Akacha M, Bretz F, Ruberg SJ. Estimands in clinical trials – broadening the perspective. *Stat Med*. 2017;36(1):5–19. doi:10.1002/sim.7033.

Alosh M, Fritsch K, Huque M, et al. Statistical considerations on subgroup analysis in clinical trials. *Stat Biopharm Res*. 2015;7(4):286–303. doi:10.1080/19466315.2015.1077726.

Angrist JD, Imbens GW, Rubin DB. Identification of causal effects using instrumental variables. *J Am Stat Assoc*. 1996;91:444–455.

Barnard J, Frangakis CE, Constantine E, Hill JL, Rubin DB. Principal stratification approach to broken randomized experiments. *J Am Stat Assoc*. 2003;98:299–323. doi:10.1198/016214503000071.

Bleecker ER, FitzGerald JM, Chanez P, et al. Efficacy and safety of benralizumab for patients with severe asthma uncontrolled with high-dosage inhaled corticosteroids and long-acting β2-agonists (SIROCCO): A randomised, multicentre, placebo-controlled phase 3 trial. *The Lancet*. 2016;388(10056):2115–2127.

Bornkamp B, Bermann G. Estimating the treatment effect in a subgroup defined by an early post-baseline biomarker measurement in randomized clinical trials with time-to-event endpoint. *Stat Biopharm Res*. 2019. doi:10.1080/19466315.2019.1575280.

Busse WW, Pedersen S, Pauwels RA, Tan WC, Chen YZ, Lamm CJ, O'byrne PM. The inhaled steroid treatment as regular therapy in early asthma (START) study 5-year follow-up: Effectiveness of early intervention with budesonide in mild persistent asthma. *J Allergy Clin Immunol*. 2008;121(5):1167–1174.

Cao W, Tsiatis AA, and Davidian M. Improving efficiency and robustness of the doubly robust estimator for a population mean with incomplete data. *Biometrika* 2009;96:723–734.

Carpenter J, Roger J, Kenward M. Analysis of longitudinal trials with missing data: A framework for relevant, accessible assumptions, and inference via multiple imputation. *J Bio Pharm Stat*. 2013;23:1352–1371.

Carpenter JR, Kenward MG. (2007). *Missing Data in Clinical Trials – A Practical Guide*. Birmingham: National Health Service Coordinating Centre for Research Methodology. Available at http://www.pcpoh.bham.ac.uk/publichealth/methodology/projects/RM03JH17MK.shtml. Accessed May 28, 2009.

Carpenter JR, Kenward MG. (2013). *Multiple Imputation and Its Application*. Chichester, UK: John Wiley & Sons. doi:10.1002/9781119942283.

Carpenter JR, Kenward MG, Vansteelandt S. A comparison of multiple imputation and doubly robust estimation for analyses with missing data. *J Royal Stat Soc A*. 2006;169:571–584.

Colantuoni E, Scharfstein DO, Wang C, et al. Statistical methods to compare functional outcomes in randomized controlled trials with high mortality. *BMJ*. 2018;360:j5748.

Chiba Y, VanderWeele TJ. A simple method for principal strata effects when the outcome has been truncated due to death. *Am J Epidem*. 2011;173:745–751.

Collins LM, Schafer JL, Kam CM. A comparison of inclusive and restrictive strategies in modern missing data procedures. *Psychol Methods* 2001;6(4):330–351.

Colmegna I, Ohata BR, Menard HA. Current understanding of rheumatoid arthritis therapy. *Clin Pharmacol Ther.* 2012;91:607–620.

Cro S, Carpenter JR, Kenward MG. Information-anchored sensitivity analysis: Theory and application. *J Royal Stat Soc A.* 2019;182:623–645.

Crowe B, Lipkovich I, Wang O. Comparison of several imputation methods for missing baseline data in propensity scores analysis of binary outcome. *Pharm Stat.* 2009;9:269–279.

Crump SL. The Estimation of Variance in Multiple Classification, PhD thesis, Department of Statistics, Iowa State University, 1974.

Daniel RM, Cousens SN, De Stavola BL, Kenward MG, Sterne JA. Methods for dealing with time-dependent confounding. *Stat Med.* 2013;32:1584–1618.

Daniel RM, Kenward MG. A method for increasing the robustness of multiple imputation. *Comput Stat Data Anal.* 2012;56(6):1624–1643.

Daniels MJ, Hogan JW. (2008). *Missing Data in Longitudinal Studies.* Boca Raton, FL: Chapman & Hall/CRC Press.

Dawid AP. Causal inference without counterfactuals (with discussion and rejoinder). *J Am Stat Assoc.* 2000;95:407–424.

Detke MJ, Wiltse CG, Mallinckrodt CH, McNamara RK, Demitrack MA, Bitter I. Duloxetine in the acute and long-term treatment of major depressive disorder: A placebo- and paroxetine-controlled trial. *Eur Neuropsychopharm.* 2004;14(6):457–470.

Diagnosing and Treating Depression – Adult – Primary Care Clinical Practice Guideline; 2010. Available at https://www.pplusic.com/documents/upload/guidelines-depression.pdf.

Diggle P, Heagerty P, Liang K-Y, Zeger S. (2002). *Analysis of Longitudinal Data.* Oxford Statistical Science Series. 2nd ed. Oxford: Oxford University Press.

Ding P, Lu J. Principal stratification analysis using principal scores. *J Royal Stat Soc B.* 2017;79(3):757–777.

European Medicines Agency (EMA), Committee for Medicinal Products for Human Use (CHMP). Guideline on the Clinical Investigation of Medicinal Products for the Treatment of Asthma. Version October 22, 2015. Available at http://www.ema.europa.eu.

European Medicines Agency (EMA), Committee for Medicinal Products for Human Use (CHMP). Fasenra Assessment Report; 2017. Available at http://www.ema.europa.eu/docs/en_GB/document_library/EPAR_-_Public_assessment_report/human/004433/WC500245333.pdf.

Felson DT, Smolen JS, Wells G, et al. American college of rheumatology/European league against rheumatism provisional definition of remission in rheumatoid arthritis for clinical trials. *Ann Rheum Dis.* 2011;70:404–413.

Firth D. Bias reduction of maximum likelihood estimates. *Biometrika.* 1993;80:27–38.

Fitzmaurice GM, Laird NM, Ware JH. (2004). *Applied Longitudinal Analysis.* New York: John Wiley & Sons.

Fitzmaurice GM, Molenberghs G, Lipsitz SR. Regression models for longitudinal binary responses with informative dropouts. *J Royal Stat Soc B.* 1995;57:691–704.

Fleming TR. Addressing missing data in clinical trials. *Ann Intern Med.* 2011;154:113–117.

Food and Drug Administration (FDA), Center for Drug Evaluation and Research. Application Number: 202293Orig1s000, Summary Review(s); 2014a. Available at https://www.accessdata.fda.gov/drugsatfda_docs/nda/2014/202293Orig1s000SumR.pdf.

Food and Drug Administration (FDA), Center for Drug Evaluation and Research. Application Number: 202293Orig1s000, Medical Review(s); 2014b. Available at https://www.accessdata.fda.gov/drugsatfda_docs/nda/2014/202293Orig1s000MedR.pdf.

Food and Drug Administration (FDA), Center for Drug Evaluation and Research. Application Number: 202293Orig1s000, Statistical Review(s); 2014c. Available at https://www.accessdata.fda.gov/drugsatfda_docs/nda/2014/202293Orig1s000StatR.pdf.

Food and Drug Administration (FDA), Center for Drug Evaluation and Research. Application Number: 761070Orig1s000, Clinical Review(s); 2017a. Available at https://www.accessdata.fda.gov/drugsatfda_docs/nda/2017/761070Orig1s000MedR.pdf.

Food and Drug Administration (FDA), Center for Drug Evaluation and Research. Application Number: 761070Orig1s000, Statistical Review(s); 2017b. Available at https://www.accessdata.fda.gov/drugsatfda_docs/nda/2017/761070Orig1s000StatR.pdf.

Food and Drug Administration (FDA). Guidance for Industry: Rheumatoid arthritis: Developing drug products for treatment, draft, revision 1, May 2013. Available at https://www.fda.gov/downloads/drugs/guidancecom.

Food and Drug Administration (FDA). Major Depressive Disorder: Developing drugs for treatment. Guidance for industry, draft, revision 1, June 2018. Available at https://www.fda.gov/downloads/Drugs/GuidanceCompliance RegulatoryInformation/Guidances/UCM611259.pdf.

Frangakis CE, Rubin DB. Principal stratification in causal inference. *Biometrics*. 2002;58(1):21–29. doi:10.1111/j.0006-341X.2002.00021.x.

Garrett A. Choosing appropriate estimands in clinical trials (Leuchs et al): Letter to the editor. *Ther Innov Regul Sci*. 2015. doi: 10.1177/21684790155860.

Global Asthma Network. The Global Asthma Report; 2014. Available at http://globalasthmareport.org.

Global Asthma Network. The Global Asthma Report; 2018. Available at http://globalasthmareport.org.

Global Initiative for Asthma. Global Strategy for Asthma Management and Prevention; 2018. Available at www.ginasthma.org.

Goldstein DJ, Mallinckrodt CH, Lu Y, Demitrack MA. Duloxetine in the treatment of major depression: A double-blind clinical trial. *J Clin Psychiat*. 2002;63(3):225–231.

Gould AL. A new approach to the analysis of clinical drug trial with withdrawals. *Biometrics*. 1980;36:721–727.

Graham JW, Olchowski AE, Gilreath TD. How many imputations are really needed? Some practical clarifications of multiple imputation theory. *Prev Sci*. 2007;8:206–213.

Greenland S, Pearl J, Robins JM. Causal diagrams for epidemiological research. *Epidemiology*. 1999;10(1):37–48.

Guy W. (1976). ECDEU Assessment *Manual for Psychopharmacology*. Rockville, MD: National Institute of Mental Health, pp. 217–222, 313–331.

Hamilton M. A rating scale for depression. *J Neurol Neurosurg Psychiat*. 1960;23:56–61.

Hammer SM, Katzenstein DA, Hughes MD, et al. A trial comparing nucleoside monotherapy with combination therapy in HIV-infected adults with cd4 cell counts from 200 to 500 per cubic millimeter. *J Med*. 1996;335(15):1081–1090.

Hartley HO, Rao JNK. Maximum-likelihood estimation for the mixed analysis of variance model. *Biometrika*. 1967;54:93–108.

Harville DA. Maximum likelihood approaches to variance component estimation and to related problems. *J Am Stat Assoc.* 1977;72, 320–338.

Heitjan DF, Rubin DB. Ignorability and coarse data. *Ann Stat.* 1991;19:2244–2253.

Hernán MA, Brumback BA, Robins JM. Marginal structural models to estimate the causal effect of zidovudine on the survival of HIV-positive men. *Epidemiology.* 2000;11(5):561–570.

Hoffman J, van Griensven J, Colebunders R, McKellar M. Role of the CD4 count in HIV management. *HIV Ther.* 2010;4(1):27–39.

Hughes S, Harris J, Flack N, Cuffe RL. The statistician's role in the prevention of missing data. *Pharm Stat.* 2012;11(5):410–416. doi: 10.1002/pst.1528.

Horvitz DG, Thompson DJ. A generalization of sampling without replacement from a finite universe. *J Am Stat Assoc.* 1952;47(260):663–685.

Hurvich CM, Tsai C-L. Regression and time series model selection in small samples. *Biometrika.* 1989;76:297–307.

ICH Harmonised Guideline E9. Statistical principles for clinical trials. Step 5 version dated September 1998.

ICH Harmonised Guideline E9(R1). Estimands and sensitivity analysis in clinical trials. Step 4 version dated November 20, 2019.

ICH Harmonised Guideline E10. Choice of control group and related issues in clinical trials. Step 4 version dated July 20, 2000. Available at https://database.ich.org/sites/default/files/E10_Guideline.pdf.

Imbens G, Angrist J. Identification and estimation of average treatment effects. *Econometrica.* 1994;62:467–475.

Imbens GW, Rubin DB. Bayesian inference for causal effects in randomized experiments with noncompliance. *Ann Stat.* 1997;25:305–327.

Jansen I, Beunckens C, Molenberghs G, Verbeke G, Mallinckrodt CH. Analyzing incomplete binary longitudinal clinical trial data. *Stat Sci.* 2006;21(1):52–69.

Jennrich RI, Schluchter MD. Unbalanced repeated measures models with structural covariance matrices. *Biometrics.* 1986;42(4):805–820.

Jo B, Stuart EA. On the use of propensity scores in principal causal effect estimation. *Stat Med.* 2009;28(23):2857–2875.

Johnson BA, Tsiatis AA. Semiparametric inference in observational duration-response studies, with duration possibly right-censored. *Biometrika.* 2005;92:605–618.

Kang JDY, Schafer JL. Demystifying double robustness: A comparison of alternative strategies for estimating a population mean from incomplete data (with discussion and rejoinder). *Stat Sci.* 2007;22(4):523–539.

Kawaguchi A, Koch G. Sanon: An R package for stratified analysis with nonparametric covariable adjustment. *J Stat Softw.* 2015;67(9). doi:10.18637/jss.v067.i09.

Keene ON, Roger JH, Hartley BF, Kenward MG. Missing data sensitivity analysis for recurrent event data using controlled imputation. *Pharm Stat.* 2014;13:258–264.

Kenward, MG, Molenberghs G. Last observation carried forward: A crystal ball? *J Biopharm Stat.* 2009;19:872–888.

Kessler RC, Berglund P, Demler O, Jin R, Merikangas KR, Walters EE. Lifetime prevalence and age-of-onset distributions of DSM-IV disorders in the national comorbidity survey replication. *Arch Gen Psychiat.* 2005;62(6):593–602.

Keystone EC, Genovese MC, Klareskog L, et al. Golimumab, a human antibody to tumour necrosis factor α given by monthly subcutaneous injections, in active rheumatoid arthritis despite methotrexate therapy: The GO-FORWARD Study. *Ann Rheum Dis.* 2009;68:789–796.

Klareskog L, Catrina AI, Paget S. Rheumatoid arthritis. *Lancet*. 2009;373(9664):659–672.

Lachin JM. Worst-rank score analysis with informatively missing observations in clinical trials. *Control Clin Trials*. 1999;20:408–422. doi:10.1016/S0197-2456(99)00022-7.

Laird NM, Ware JH. Random-effects models for longitudinal data. *Biometrics*. 1982;38:963–974.

Leuchs AK, Zinserling J, Brandt A, Wirtz D, Benda N. Choosing appropriate estimands in clinical trials. *Ther Innov Regul Sci*. 2015; 49(4):584–592. doi:10.1177/2168479014567317.

Liang KY, Zeger SL. Longitudinal data analysis using generalized linear models. *Biometrika*. 1986;73:13–22.

Lindstrom MJ, Bates DM. Newton-raphson and EM algorithms for linear mixed-effects models for repeated-measures data. *J Am Stat Assoc*. 1988;83:1014–1022.

Lipkovich I, Kadziola Z, Xu L, Sugiharad T, Mallinckrodt CH. Comparison of several multiple imputation strategies for repeated measures analysis of clinical scales: Truncate or not to? *J Biopharm Stat*. 2014;24:924–943.

Lipkovich I, Ratitch B, Mallinckrodt CH. Causal inference and estimands in clinical trials. *Stat Biopharm Res*. 2019. doi:10.1080/19466315.2019.1697739.

Lipkovich I, Ratitch B, O'Kelly M. Sensitivity to censored-at-random assumption in the analysis of time-to-event endpoints. *Pharm Stat*. 2016;15(3):216–229.

Lipkovich I, Wiens B. The role of multiple imputation in noninferiority trials for binary outcomes. *Stat Biopharm Res*. 2018;10(1):57–69.

Little R, Kang S. Intention-to-treat analysis with treatment discontinuation and missing data in clinical trials. *Stat Med*. 2014;34:2381–2390.

Little RJ, Rubin DB. Causal effects in clinical and epidemiological studies via potential outcomes. *Ann Rev Publ Health*. 2000;21:121–145.

Little R, Yau L. Intent-to-treat analysis for longitudinal studies with dropouts. *Biometrics*. 1996;52(4):1324–1333.

Little RJA. Modeling the drop-out mechanism in repeated measures studies. *J Am Stat Assoc*. 1995;90(431):1112–1121.

Little RJA, Rubin DB. (1987). *Statistical Analysis with Missing Data*. 1st ed. New York: Wiley.

Little RJA, Rubin DB. (2002). *Statistical Analysis with Missing Data*. 2nd ed. New York: Wiley.

Liu GF, Lu K, Mogg R, Mallick M, Mehrotra DV. Should baseline be a covariate or dependent variable in analyses of change from baseline in clinical trials? *Stat Med*. 2009;28:2509–2530.

Liu GF, Pang L. On analysis of longitudinal clinical trials with missing data using reference-based imputation. *J Biopharm Stat*. 2016;5:924–936. doi:10.1080/10543406.2015.1094810.

Lu X, Mehrotra DV, Shepherd BE. Rank-based principal stratum sensitivity analyses. *Stat Med*. 2013;32(26):4526–4539.

Magnusson BP, Schmidli H, Rouyrre N, Scharfstein DO. Bayesian inference for a principal stratum estimand to assess the treatment effect in a subgroup characterized by post-randomization events. *Stat Med*. 2019;38:4761–4771.

Mallinckrodt C, Bell J, Liu G, Ratitch B, O'Kelly M, Lipkovich I, Singh P, Xu L, Molenberghs G. Technical and practical considerations in aligning estimators with estimands in clinical trials. *Ther Innov Regul Sci*. doi:10.1177/2168479019836979.

Mallinckrodt CH. (2013). *Preventing and Treating Missing Data in Longitudinal Clinical Trials: A Practical Guide*. New York: Cambridge University Press.

Mallinckrodt CH, Clark WS, Carroll RJ, Molenberghs G. Assessing response profiles from incomplete longitudinal clinical trial data under regulatory conditions. *J BioPharm Stat.* 2003;13(2):179–190.

Mallinckrodt CH, Lane PW, Schnell D, Peng U, Mancuso JP. Recommendations for the primary analysis of continuous endpoints in longitudinal clinical trials. *Drug Inf J.* 2008;42:305–319.

Mallinckrodt CH, Lin Q, Molenberghs M. A structured framework for assessing sensitivity to missing data assumptions in longitudinal clinical trials. *Pharm Stat.* 2013;12:1–6.

Mallinckrodt CH, Lin Q, Lipkovich I, Molenberghs G. A structured approach to choosing estimands and estimators in longitudinal clinical trials. *Pharm Stat.* 2012;11:456–461.

Mallinckrodt CH, Lipkovich I. *A Practical Guide to Analyzing Longitudinal Clinical Trial Data.* Boca Raton, FL: Chapman & Hall/CRC.

Mallinckrodt CH, Molenberghs G, Rathmann S. Choosing estimands in clinical trials with missing data. *Pharm Stat.* 2017;16(1):29–36. doi:10.1002/pst.1765.

Mallinckrodt CH, Roger J, Chuang-Stein C, et al. Recent developments in the prevention and treatment of missing data. *Ther Innov Regul Sci.* 2014;48(1):68–80.

McCullagh P, Nelder JA. (1989). *Generalized Linear Models.* 2nd ed. London, UK: Chapman and Hall.

Mehrotra D, Li X, Gilbert PA. Comparison of eight methods for the dual-endpoint evaluation of efficacy in a proof-of-concept HIV vaccine trial. *Biometrics.* 2006;62:893–900.

Mehrotra D, Liu F, Permutt T. Missing data in clinical trials: Control-based mean imputation and sensitivity analyses. *Pharm Stat.* 2017;16:378–392.

Meng XL. Multiple imputation with uncongenial sources of input (with discussions). *Stat Sci.* 1994;9:538–574.

Miratrix L, Furey J, Feller A, Grindal T, Page LC. Bounding, an accessible method for estimating principal causal effects, examined and explained. *J Res Educ Eff.* 2018;11(1):133–162.

Molenberghs G, Fitzmaurice G, Kenward MG, Verbeke G, Tsiatis AA. (2015). *Handbook of Missing Data.* Boca Raton, FL: Chapman & Hall/CRC.

Molenberghs G, Kenward MG. (2007). *Missing Data in Clinical Studies.* Chichester, UK: John Wiley & Sons.

Molenberghs G, Verbeke G. (2005). *Models for Discrete Longitudinal Data.* New York: Springer.

Moodie EEM, Richardson TS, Stephens DA. Demystifying optimal dynamic treatment regimes. *Biometrics.* 2007;63:447–455.

Murphy SA. An experimental design for the development of adaptive treatment strategies. *Stat Med.* 2005;24(10):1455–1481.

Murphy SA. Optimal dynamic treatment regimes (with discussion). *J Royal Stat Soc.* 2003;65(2):331–366.

National Research Council (NRC). (2010). *The prevention and Treatment of Missing Data in Clinical Trials.* Panel on Handling Missing Data in Clinical Trials. Committee on National Statistics, Division of Behavioural and Social Sciences and Education. Washington, DC: The National Academies Press.

Neyman J. On the application of probability theory to agricultural experiments. Essay on principles. Section 9, *Stat Sci.* 1923;5:465–480.

O'Kelly M, Ratitch B. (2014). *Clinical Trials with Missing Data.* Chichester, UK: Wiley.

O'Neill RT, Temple R. The prevention and treatment of missing data in clinical trials: An FDA perspective on the importance of dealing with it. *Clin Pharmacol Ther.* 2012. doi:10.1038/clpt.2011.340.

Patterson HD, Thompson R. Recovery of inter-block information when block sizes are unequal. *Biometrika.* 1971;58:545–554.

Pearl J. Causal inference in the Health Sciences: A conceptual introduction. *Health Serv Outcomes Res Methodol.* 2001;2:189–220.

Permutt T. A taxonomy of estimands for regulatory clinical trials with discontinuations. *Stat Med.* 2015a. doi:10.1002/sim.6841.

Permutt T. Sensitivity analysis for missing data in regulatory submissions. *Stat Med.* 2015b. doi:10.1002/sim.6753.

Permutt T, Li F. Trimmed means for symptom trials with dropouts. *Pharm Stat.* 2017;16:20–28.

Phillips A, Abellan-Andres J, Andersen S. et al. Estimands: Discussion points from the PSI estimands and sensitivity expert group. *Pharm Stat.* 2016. doi:10.1002/pst.1745.

Phillips A, Abellan-Andres J, Soren A, et al. Estimands: Discussion points from the PSI estimands and sensitivity expert group. *Pharm Stat.* 2017;16(1):6–11.

Potter, WZ, Mallinckrodt CH, Detke MJ. Controlling placebo response in drug development: Lessons from psychopharmacology. *Pharm Med.* 2014;28:53–65. doi:10.1007/s40290-014-0052-8.

Radner H, Alasti F, Smolen JS, Aletaha D. Physical function continues to improve when clinical remission is sustained in rheumatoid arthritis patients. *Arthritis Res Ther.* 2015;17:203.

Ratitch B, Bell J, Mallinckrodt C, Bartlett J, Goele N, Molenberghs G, O'Kelly M, Singh P, Lipkovich I. Choosing estimands in clinical trials: Putting the ICH E9(R1) into practice. *Ther Innov Regul Sci.* 2019a. doi:10.1177/2168479019838827.

Ratitch B, Goel N, Mallinckrodt C, Bell J, Bartlett J, Molenberghs G, Singh P, Lipkovich I, O'Kelly M. Defining estimands in clinical trials: Examples illustrating ICH E9(R1) guidelines. *Ther Innov Regul Sci.* 2019b. doi: 10.1177/2168479019841316.

Ratitch B, Lipkovich I, Erickson JS, Zhang L, Mallinckrodt CH. Points to consider for analyzing efficacy outcomes in long-term extension clinical trials. *Pharm Stat.* 2018;17:685–700.

Ratitch B, O'Kelly M, Tosiello R. Missing data in clinical trials: From clinical assumptions to statistical analysis using pattern mixture models. *Pharm Stat.* 2013;12(6):337–347. doi:10.1002/pst.1549.

Robins JM. A new approach to causal inference in mortality studies with a sustained exposure period – application to control of the healthy worker survivor effect. *Math Model.* 1986;7:1393–1512.

Robins JM. Correction for non-compliance in equivalence trials. *Stat Med.* 1998;17:269–302.

Robins JM, Hernán MA. (2009). Estimation of the causal effects of time-varying exposures. In: *Longitudinal Data Analysis*, eds. G. Fitzmaurice, M. Davidian, G. Verbeke, and G. Molenberghs, New York, NY: Chapman and Hall/CRC Press, pp. 553–599.

Robins JM, Hernán MA, Brumback B. Marginal structural models and causal inference in epidemiology. *Epidemiology.* 2000;11(5):550–560.

Robins JM, Rotnitzky A. Semiparametric efficiency in multivariate regression models with missing data. *J Am Stat Assoc.* 1995;90:122–129.

Robins JM, Rotnitzky A, Zhao LP. Analysis of semiparametric regression models for repeated outcomes in the presence of missing data. *J Am Stat Assoc.* 1995;90:106–121.

Robins JM, Wang N. Inference for imputation estimators. *Biometrika.* 2000;87:113–124.

Roy J, Hogan JW, Marcus BH. Principal stratification with predictors of compliance for randomized trials with 2 active treatments. *Biostatistics.* 2008;9(2):277–289. doi:10.1093/biostatistics/kxm027.

Rubin DB. Comment on causal inference without counterfactuals. *J Am Stat Assoc.* 2000;95(450):435–438.

Rubin DB. (1978a). Multiple imputations in sample surveys – a phenomenological Bayesian approach to nonresponse. In: *Imputation and Editing of Faulty or Missing Survey Data.* Washington, DC: U.S. Department of Commerce, pp. 1–23.

Rubin DB. Bayesian inference for causal effects: The role of randomization. *Ann Stat.* 1978b;6:34–58.

Rubin DB. Inference and missing data. *Biometrika.* 1976;63(3):581–592.

Rubin DB. More powerful randomization-based p-values in double-blind trials with noncompliance (with discussion) *Stat Med.* 1998;17:371–389.

Rubin DB. (1987). *Multiple Imputation for Nonresponse in Surveys.* New York: John Wiley & Sons.

SAS Institute Inc. (2013). *SAS/STAT® 9.4 User's Guide.* Cary, NC: SAS Institute.

Schafer JL. (1997). *Analysis of Incomplete Multivariate Data.* New York: Chapman and Hall.

Schomaker M, Heumann C. Bootstrap Inference When Using Multiple Imputation; 2016. Available at https://arxiv. org/abs/1602.07933.

Schwartz S, Li F, Reiter R. Sensitivity analysis for unmeasured confounding in principal stratification setting with binary variables. *Stat Med.* 2012;31(10):949–962.

Seaman S, Copas A. Doubly robust generalized estimating equations for longitudinal data. *Stat Med.* 2009;28:937–955.

Shepherd BE, Gilbert PB, Dupont CT. 2011. Sensitivity analyses comparing time-to-event outcomes only existing in a subset selected postrandomization and relaxing monotonicity. *Biometrics.* 2011;67(3):1100–1110. doi:10.1111/j.1541-0420. 2010.01508.x.

Shih WJ. Problems in dealing with missing data and informative censoring in clinical trials. *Curr Control Trials Cardiovasc Med.* 2002;3:4.

Siddiqui O, Hung HM, O'Neill RO. MMRM vs. LOCF: A comprehensive comparison based on simulation study and 25 NDA datasets. *J BioPharm Stat.* 2009;19(2):227–246.

Smolen J, Steiner G. Therapeutic strategies for rheumatoid arthritis. *Nat Rev Drug Discov.* 2003;2:473–488.

Smolen JS, Breedveld FC, Burmester GR, et al. Treating rheumatoid arthritis to target: 2014 update of the recommendations of an international task force. *Ann Rheum Dis.* 2016;75:3–15.

Sullivan TR, White IR, Salter AB, Ryan P, Lee KJ. Should multiple imputation be the method of choice for handling missing data in randomized trials? *Stat Methods Med Res.* 2018;27(9):2610–2626.

Tanner MA, Wong WH. The calculation of posterior distributions by data augmentation. *J Am Stat Assoc.* 1987;82:528–540.

Thompson WA. (1962). The problem of negative estimates of variance components. *J Am Stat Assoc.* 1962;33:273–289.

Tsiatis AA. (2006). *Semiparametric Theory and Missing Data*. New York: Springer.

Tsiatis AA, Davidian M, Cao W. Improved doubly robust estimation when data are monotonely coarsened, with application to longitudinal studies with dropout. *Biometrics*. 2011;67:536–545.

van Buuren S. Multiple imputation of discrete and continuous data by fully conditional specification. *Stat Methods Med Res*. 2007;16:219–242.

van Buuren S. (2018). *Flexible Imputation of Missing Data*. 2nd ed. Boca Raton, FL: Chapman & Hall/CRC.

VanderWeele TJ. Principal stratification – Uses and limitations. *Int J Biostat*. 2011;7(1), 1–14. Article 28.

VanderWeele TJ, Hernán MA. Causal inference under multiple versions of treatment. *J Causal Inference*. 2013;1(1):1–20.

VanderWeele TJ, Vansteelandt S. Conceptual issues concerning mediation, interventions and composition. *Stat Interface* 2009;2:457–468.

Vansteelandt S, Carpenter J, Kenward MG. Analysis of incomplete data using inverse probability weighting and doubly robust estimators. *Methodology*. 2010;6(1):37–48.

Verbeke G, Molenberghs G. (2000). *Linear Mixed Models for Longitudinal Data*. New York: Springer.

Wang D, Pocock S. A win ratio approach to comparing continuous non-normal outcomes in clinical trials. *Pharm Stat*. 2016;15:238–245.

Wang MD, Liu J, Molenberghs G, Mallinckrodt CH. An evaluation of the trimmed mean approach in clinical trials with dropout. *Pharm Stat*. 2018;17(3):278–289.

Wang N, Robins JM. Large-sample theory for parametric multiple imputation procedures. *Biometrika*. 1998;85:935–948.

Wonnacott TH, Wonnacott RJ. (1981). *Regression: A Second Course in Statistics*. New York: Wiley.

Work Group on Major Depressive Disorder. (2010). *Practice Guideline for the Treatment of Patients with Major Depressive Disorder*. 3rd ed. Available at https://psychiatry-online.org/pb/assets/raw/sitewide/practice_guidelines/guidelines/mdd.pdf.

Zhang JL. Rubin DB. Estimation of causal effects via principal stratification when some outcomes are truncated by "death". *J Educ Behav Stat*. 2003;28:353–368. doi:10.3102/10769986028004353.

Glossary

estimand: is the target of estimation to address the scientific question of interest posed by the trial objective. Attributes of an estimand include the population of interest, the variable (or endpoint) of interest, the specification of how intercurrent events are reflected in the scientific question of interest, and the population-level summary for the variable.

estimate: is the numerical value computed by an estimator based on the observed clinical trial data.

estimator: is the analytic approach to compute an estimate from the observed clinical trial data.

intercurrent events: events that occur after treatment initiation and either preclude observation of the variable or affect its interpretation.

missing data: data that would be meaningful for the analysis of a given estimand but were not collected. They should be distinguished from data that does not exist or data that is not considered meaningful because of an intercurrent event.

principal stratification: is the classification of patients according to the potential occurrence of an intercurrent event on all treatments. With two treatments, there are four principal strata with respect to a given intercurrent event: patients who would not experience the event on either treatment, patients who would experience the event on treatment A but not B, patients who would experience the event on treatment B but not A, and patients who would experience the event on both treatments.

principal stratum: is used in this document to refer to any of the strata (or the combination of strata) defined by principal stratification.

sensitivity analysis: is a series of analyses targeting the same estimand, with differing assumptions to explore the robustness of inferences from the main estimator to deviations from its underlying modeling assumptions and limitations in the data.

supplementary analysis: is a general description for analyses that are conducted in addition to the main and sensitivity analysis to provide additional insights into the understanding of the treatment effect. The term describes a broader class of analyses than sensitivity analyses.

Index

Note: Page numbers in italic and bold refer to figures and tables, respectively.

Printed in the United States
by Baker & Taylor Publisher Services